# Studies in Logic
Volume 113

# Meaning as a Set-theoretic Object
A Gentle Introduction to the Ideas Behind Formal Semantics

Volume 103
The Fertile Debate. Affective Exploration of a Controversy
Claire Polo

Volume 104
Argument, Sex and Logic
Dov Gabbay, Gadi Rozenberg and Lydia Rivlin

Volume 105
Logic as a Tool. A Guide to Formal Logical Reasoning
Valentin Goranko

Volume 106
New Directions in Term Logic
George Englebretsen, ed

Volume 107
Non-commutative Algebras. Pseudo-BCK Algebreas versus m-pseudo-BCK Algebras
Afrodita Iorgulescu

Volume 108
Semitopology: decentralised collaborative action via topology, algebra, and logic
Murdoch J. Gabbay

Volume 109
The Cognitive Dimension of Social Argumentation. Proceedings of the 4th European Conference on Argumentation, Volume I. Fabio Paglieri, Alessandro Ansani and Marco Marini, eds.

Volume 110
The Cognitive Dimension of Social Argumentation. Proceedings of the 4th European Conference on Argumentation, Volume II. Fabio Paglieri, Alessandro Ansani and Marco Marini, eds.

Volume 111
The Cognitive Dimension of Social Argumentation. Proceedings of the 4th European Conference on Argumentation, Volume III. Fabio Paglieri, Alessandro Ansani and Marco Marini, eds.

Volume 112
Implicative-groups vs. Groups and Generalizations. Second Edition
Afrodita Iorgulescu

Volume 113
Meaning as a Set-theoretic Object. A Gentle Introduction to the Ideas Behind Formal Semantics
Jaroslav Peregrin

Studies in Logic Series Editor
Dov Gabbay                                              dov.gabbay@kcl.ac.uk

# Meaning as a Set-theoretic Object
A Gentle Introduction to the Ideas Behind Formal Semantics

Jaroslav Peregrin

Institue of Philosophy

Czech Academy of Sciences

© Institute of Philosophy, Czech Academy of Sciences and College Publications, 2025
All rights reserved.

This book was written as part of the grant project 23-07119S "Meaning as an object" supported by the Czech Science Foundation and coordinated by the Institute of Philosophy of the Czech Academy of Sciences in Prague.

ISBN 978-1-84890-491-0

College Publications
Scientific Director: Dov Gabbay
Managing Director: Jane Spurr

http://www.collegepublications.co.uk

Cover prepared by Debbie Hunt

All rights reserved. No part of this publication may be reproduced, stored in a retrieval system or transmitted in any form, or by any means, electronic, mechanical, photocopying, recording or otherwise without prior permission, in writing, from the publisher.

# Table of Contents

## 1 MEANING AS AN OBJECT ... 1
### 1.1 Formal semantics ... 1
### 1.2 Frege on ideal objects ... 4
### 1.3 Sets ... 7
### 1.4 Frege's maneuver ... 9
### 1.5 Semantic models of language ... 12
### 1.6 The four kinds of models ... 14

## 2 A VERY SHORT HISTORY OF FORMAL SEMANTICS ... 17
### 2.1 Frege ... 17
### 2.2 Tarski ... 22
### 2.3 Carnap ... 26
### 2.4 Standard Logic and its Semantics ... 29
### 2.5 Chomsky ... 32
### 2.6 Montague and since ... 36

## 3 EXTENSIONAL MODEL OF MEANING: FREGE'S MANEUVER EXPLOITED TO THE BONE ... 39
### 3.1 Principles of etensional semantics ... 39
### 3.2 Subject and predicate ... 41
### 3.3 Functions ... 46
### 3.4 Quantification ... 48
### 3.5 Negation and propositional connectives ... 51
### 3.6 Basic extensional model ... 55
### 3.7 Fregean quantification ... 57
### 3.8 Predicates of greater arities ... 64
### 3.9 Modified extensional model ... 66
### 3.10 Categorial grammar ... 68
### 3.11 Type theory ... 71
### 3.12 Lambda abstraction ... 73
### 3.13 Lambda-categorial grammar ... 78
### 3.14 Forging functions ... 80
### 3.15 Generalized quantifiers ... 81

## 4 INTENSIONAL MODEL OF MEANING: POSSIBLE WORLDS .......... 85

    4.1 Limits of extensional semantics .......... 85
    4.2 Modal logic and the concept of possible world .......... 88
    4.3 Extensions vs. intensions .......... 92
    4.4 Kripkean semantics .......... 94
    4.5 Temporal logic .......... 96
    4.6 From dependency+ to function .......... 97
    4.7 Possible worlds and individuals .......... 98
    4.8 Montague's grammar and locally intensional logic .......... 101
    4.9 Montague's approach to intensions .......... 108
    4.10 Globally intensional logic .......... 109
    4.11 Two-sorted type theory .......... 111

## 5 HYPERINTENSIONAL MODELS OF MEANING: STRUCTURE INCORPORATED .......... 115

    5.1 Propositional attitudes and intensional isomorphism .......... 115
    5.2 The "semantic structure" of an expression .......... 118
    5.3 Structured meanings .......... 122
    5.4 Are meanings still sets? .......... 125
    5.5 Tichý's constructions .......... 126
    5.6 Situation semantics .......... 130
    5.7 Situations vs. possible worlds .......... 134
    5.8 From extension to situation .......... 138
    5.9 Situation and mental representation .......... 142
    5.10 Where are meanings? .......... 143
    5.11 Autonomy of semantic structure .......... 145

## 6 DYNAMIC MODELS OF MEANINGS: CONTEXTS AND UTTERANCES .......... 149

    6.1 Problems of anaphoric reference .......... 149
    6.2 Articles .......... 152
    6.3 Discourse representation theory .......... 155
    6.4 Meaning as a change of state .......... 161
    6.5 Information states and their updates .......... 163
    6.6 Information states as situations and as sets of possible worlds .......... 166
    6.7 Dynamic predicate logic .......... 170

## 7 MEANING AS AN OBJECT .......... 175

    7.1 Formal semantics .......... 175
    7.2 Being an object vs. being conceived as an object .......... 177
    7.3 Explication .......... 179

| *7.4 | Meaning and its Formal Explication | 181 |

## 8  APPENDIX: THE MODELS .................................................. 185

| 8.1 | Basic extensional model | 185 |
| 8.2 | Modified extensional model | 188 |
| 8.3 | Categorial grammar | 190 |
| 8.4 | Lambda-categorial grammar | 191 |
| 8.5 | Locally intensional logic | 192 |
| 8.6 | Globally intensional logic | 194 |

## REFERENCES ............................................................................ 195

# 1 Meaning as an object

## 1.1 Formal semantics

Some fifty years ago, the semantics of natural language started a new era. The harbingers of the new period were thinkers operating on the boundary between logic, philosophy and linguistics. This included such figures as Montague (1974), Lewis (1972) and Cresswell (1973), who were seeking the reconstruction of meanings as set-theoretic objects, providing a very neat picture of the system of meanings reflecting (or being reflected by) the system of linguistic expressions. This raises the question as to how was all this accomplished, and whether this was a genuine breakthrough in semantics. (It is fair to note that, not by any measure, were all who were engaged in the semantics of natural language impressed by this kind of formal semantics.)

Nowadays, this breakthrough is far behind us, and we can look back at it with the benefit of hindsight. Not that the paradigm of formal semantics would fade away; analyses of various aspects and phenomena of natural language carried out within its framework keep appearing (regularly, for example, in journals like *Linguistics and Philosophy* and *Natural Language Semantics*). But the heated debates of its conceptual foundations are long over now, and formal semantics is taken as an established (though sometimes slightly heterogeneous) framework that can provide space to discuss the semantics of natural language instead of being something to be questioned or challenged. This distance makes it possible to see the foundations of formal semantics in a clear light and to assess its merits.

The roots of this approach to meaning go back to the early days of analytic philosophy with its conviction that natural language is treacherous and that we must see through its apparent surface structures to its logical forms, which determine the contents of its sentences. The paradigmatic analysis of the sentences with definite descriptions coming from Russell (1905) tells us that a sentence like

*The king of France is bald*

does not necessarily tell us that an individual, the king of France, has a property, baldness. Rather, it tells us something much more complicated, which Russell captures by the following formula:

$$\exists x(\mathbf{KF}(x) \land \mathbf{B}(x) \land (\forall y(\mathbf{KF}(y) \to (y=x))))$$

It is only if we realize this, Russell insists, that we are clear about what the sentence really says – about its meaning. It follows, according to Russell, that the existence of the king of France is not a condition of the meaningfulness of the sentence, but just part of what the sentence asserts; if this is not the case, the sentence is simply false. This analysis also provides for the ambiguity of the negation of the sentence: while the negation sign can be put in front of the whole sentence, it can also be put in front of the predicate $\mathbf{B}$, etc.

Whatever we think about the merits of this analysis, we see that it does not tell us explicitly what the meaning of the sentence is, not to mention the meanings of its parts, such as the definite article *the*. One possible approach to a further development of this approach to language was to try to capture the meanings explicitly. And some ideas of Frege, buttressed by the development of set theory, appeared to offer a way of accomplishing this. The way was negotiated, especially, by Carnap (1942; 1947), followed by Montague and company.

The outcome of this new approach was, for example, to capture the meaning of a sentence as a class of possible worlds in which the sentence is true, or to capture the meaning of the definite article as something like the function from sets to individuals which maps all singletons on their single members and all other sets on nothing. So here we have the promise of answering the question about the nature of meaning, wholly and explicitly, by presenting meanings as set-theoretical objects.

This approach to meaning has smuggled in the assumption that meanings are objects. In fact, this assumption appears quite trivial, for what else could a meaning be other than a kind of object? Imagining linguistic meanings (*viz.* that which makes a mere sound or scribble into a meaningful expression) as objects comes naturally. Is not the relationship between a word and its meaning akin to the relationship between a proper name and its barer? Yet when we start to think about words other than nominal phrases, this idea becomes ever blurrier. What kind of objects, represented by expressions, could meanings be?

Several answers spring to mind:

One possibility is that meanings are objects of the spatiotemporal world. This reinforces the parallel between proper names and linguistic expressions in general. But it seems that here we encounter what can be called *the problem of scarcity*: the spatiotemporal world does not contain entities capable of serving as meanings for all our expressions. We need not even mention words such as *always* or *notwithstanding* since for such ordinary words as *dog* or *run* there already do not seem to be single spatiotemporal objects which could be seen as their meanings. (Something like the mereological sum of all the actual dogs or runners would obviously not do.)

Another possibility is that meanings are objects of the mental realm. This appears to solve the problem of scarcity, for the mental realm can be seen as containing a bottomless fount of entities. However, here we face another kind of problem, one which can be called *the problem of intersubjectivity*: meanings are essentially intersubjective, they fulfill their role only as much as they can be shared among participants of communication, hence they cannot reside in private minds. (As Davidson, 1990, p. 314, puts it, "that meanings are decipherable is not a matter of luck; public availability is a constitutive aspect of language".)

Then there is a further possibility: meanings are objects of a realm of ideal entities. This tries to avoid both the problem of scarcity and that of intersubjectivity, for the objects of this realm are considered objective, as spatiotemporal things are, and yet are not limited by space and time. Frege (1918, p. 69) was adamant about this being the only viable possibility for meanings: "A third realm must be recognized. What belongs to this corresponds with ideas, in that it cannot be perceived by the senses, but with things, in that it needs no bearer to the contents of whose consciousness to belong."

The problem with this, however, is the status of the realm of the ideal entities. Its whereabouts seem to be slightly enigmatic and in any case its denizens seem to be causally inert, which makes it hard to explain their impact on the spatiotemporal world. But here there is an opening for sets: they are ideal entities of the above kind, and yet they are backed up by a host of respectable mathematicians, hence to call them "enigmatic" would seem inappropriate. (The causal inertness is not yet addressed, but hopefully this too could be worked around....) So perhaps sets may help us make semantics ultimately explicit. This would explain the open arms with which many theoreticians of language welcomed the set-theoretic semantics.

However, let us first return to the time before set theory came to the full function.

## 1.2 Frege on ideal objects

The general question is how we can explicitly grasp ideal objects. Frege, the ur-father of formal semantics, wrestled with this problem when he faced the question *What is a number?*, the answer to which was, in his eyes, key to understanding the foundations of mathematics. He came to the idea that such objects must be grasped in terms of their names; however, he insisted, it is not enough to have the (alleged) names of numbers, we also must have a way to determine when two names name the same number. Frege (1884, p. 73) writes: "If we are to use the symbol *a* to signify an object, we must have a criterion for deciding in all cases whether *b* is the same as *a*, even if it is not always in our power to apply this criterion. ... When we have thus acquired a means of arriving at a determinate number and of recognizing it again as the same, we can assign it a number word as its proper name."

To indicate that this approach does not concern only numbers, but abstract objects more generally, he first turns his attention, as a warm-up, to the abstract object *direction*. The direction is something that can be, in geometry, represented by a straight line, we can therefore use names such as *the direction of line a*. Hence, we have names for the directions, but in order to be able to use them as genuine names we must supply a method which determines when two of the names name the same direction. We need then a method which lets us decide, for any claim of the form *the direction of line a is the same as the direction of line b*, whether it is true. Such a statement, however, can be clearly seen as a mere paraphrase of the statement *the lines a and b are parallel*, the truth value of which we can find (at least in principle). This lets us understand directions as objects (to recognize the same direction under different names).

But all this doesn't really tell us much about what a direction *is*. But this is normal, according to Frege: "The definition of an object does not, as such, really assert anything about the object, but only lays down the meaning of a symbol" (p. 78). To constitute an abstract object, then, is to know nothing more than to establish the meaning of a certain kind of sign: we cannot establish meaning except by determining how that sign is used in certain sentences (especially certain equations), and more specifically what truth-values those sentences have. (Frege states bluntly, "It is only in the context of a proposition that words have any meaning", p. 73.) All we can say about direction is that it is something that is common to two parallel lines. Thus, the object *direction a* is identical with the object *direction b* precisely if *a* and *b* are parallel; both *direction a* and *direction b* in this case simply denote "what *a* and *b* have in common (in this respect)."

But if this is the case, then according to Frege we can identify the direction of a line without scruple with any suitable object that is associated with the line and that satisfies the condition that it is the same for two lines precisely when they are parallel. Such a convenient object, according to Frege, is in the case of a line *a* the domain of the notion of *the line parallel to a* – that is, we would say today, the set of all objects which fall under this notion, i.e. the set of parallels of *a*. For the domain of the notion *parallel to a* (the set of parallels of *a*) and the domain of the notion *parallel to b* (the set of parallels of *b*) are obviously truly identical precisely when *a* and *b* are parallel. Frege thus proposes to identify the subject *direction a* with the set of all parallels of *a*.

Does this mean that the direction is the set of parallels? (Isn't this counterintuitive? – It doesn't seem that if we were thinking of the direction of a line, we would be thinking of a set!) There is no simple answer to this question. Frege's reasoning here is essentially this: since we can say nothing more about direction than that it is what is common to all parallel lines we can pretty well identify it with anything that has this property – and disregard the fact that the direction so grasped may well take on other, non-intuitive properties. This suggestion foreshadows the method that Carnap (1947) later called *explication*: replacing some abstract and hard-to-grasp entity with something that shares all its characteristic properties and is in some sense easier to grasp, usually some mathematical construct like a set (more about it in Section 7.3).

By analogy, Frege now wants to approach the notion of number (or what he calls *Anzahl*). He states that just as direction is something that belongs to a line, number is something that belongs to a concept (so, for example, the number eight belongs to the concept of *planets of the solar system*). If, as we have already done, we call the set of all the objects falling under a concept *the domain* of that concept, we can say that the object *number of F* is identical with the object *number of G* precisely when the domain of the concept *equinumerous*[1] *to F* is identical with the domain of the concept *equinumerous to G*. In this case, then, we can – by analogy with what we have done in the case of direction – identify the number of *F* with the domain of the concept *equinumerous to F*, that is, with the set of all sets *equinumerous* to *F*.

Let us summarize the principles of Frege's approach to abstract entities. According to him, we can speak of entities of a certain kind, as we have said, if we have signs for them, and if we can say when two of these signs designate the same entity, i.e. if we have a certain equivalence between the signs. We can then look at the relevant entity as that which all such

---

[1] Frege's German term is *gleichzählig*, which is often translated as *equal* or *equivalent*. This is, I think, a little bit misleading. Two equinumerous concepts are characterized by the fact that there is a one-one mapping between their domains and, at least in the case of very small domains, we can just *see* this as we can *see* that two lines are parallel.

equivalent signs, as signs, have in common. In both of Frege's examples, moreover, the sign of the abstract entity is uniquely tied to some more "concrete" entity – the direction sign to a line and the number sign to a concept. Thus, direction can be thought of as what all parallel lines have in common, and number as what all equinumerous concepts have in common. What Frege proposes next can be understood as identifying what the elements of a set have in common with the set itself: identifying what all parallel lines have in common with the set of all parallel lines, and what all equivalent concepts have in common with the set of all equivalent concepts.

In this way, then, Frege reduces the notion of number, which is at the basis of arithmetic, to the notions of a concept and of a domain of a concept (a set), which he regarded as purely logical notions. And when the distinction between a concept and its domain was later blurred, it was possible simply to say that numbers are sets of sets.

## 1.3 Sets

What, after all, is a set? It is notoriously difficult to say – even though sets play a crucial role in the foundations of modern mathematics. The modern theory of sets originated with Georg Cantor,[2] who originally considered sets of numbers, i.e. of points on the number axis. He studied certain functions and found out that if they have certain properties at every point of their domains, something important follows; but then he realized that it keeps to follow if the functions have the properties at every point of their domain *with the exception of some of them*. (For example: it was proven, already before Cantor, that if a function is representable by a uniformly convergent trigonometric series, then the series is unique. Cantor proved not only that uniform convergence is not necessary, but that even simple convergence might fail at some exceptional points, providing the set of such points is finite or even infinite if it has a certain structure.) It turned out that he needed a theory to characterize the sets in question – hence the original version of set

---

[2] See Cantor (1932).

theory.

But the theory acquired a life of its own and started to overspill its original confines. Thus, for example, Russell took sets to derive from properties: a property, according to him, determined the set of all objects having the property. (Thus, we cannot have a set all the elements of which do not share a property.)[3] It followed that there was no reason to consider numbers as the only potential elements of sets. The property of (*being*) *hot* determines the set of all hot objects; the property of *being born in Istanbul on* 1.1.1111 determines the set of all individuals born in Istanbul on 1.1.1111, the property of *being a set of sets* determines the set of all sets whose elements are sets.

Frege's considerations presented in the previous section led Frege and Russell to conclude that numbers are nothing else than certain sets. Number five, for example, is the set of all sets with five elements (oversimplifying a little bit). This led them to the conclusion that mathematics can be embedded into logic – the view known as *logicism* (but this is a story for a different occasion).[4] And if numbers could be so easily embedded into such theory, why not other abstract objects? Functions, the abstract objects crucial for mathematics, came to be grasped as sets of ordered pairs. Thus, the function of square (of natural numbers) would be the (infinite) set $\{<1,1>, <2,4>, <3,9>, ...\}$, where $<a,b>$ is a shorthand for $\{\{a\},\{a,b\}\}$.

Hence as long as mathematics was taken to be "a science of number", the sets it was interested in were points on the number axis. But the more mathematics moved from these confines to becoming the general theory of structures, the more it could be seen as a theory of sets because all structures came to be reconstructed as set-theoretical objects. This was because set theory slowly became a general framework for reconstructing abstract and ideal objects so that reconstructibility within

---

[3] An anecdote ascribed to Bertrand Russell says that while we can have the set of all left shoes (because these can be distinguished from right ones), we cannot have that of left socks.

[4] See Demopoulos (2013).

such a theory became a hallmark of being a genuine object.

The problem with sets in mathematics was that even after sets started to acquire such an important role in the foundation of mathematics, nobody was able to say very clearly what a set is. The solution to this problem came with the axiomatizations of set theory. Several such axiomatizations were proposed, and the general feeling was that a successful axiomatization gives us the answer to the question about the nature of sets. True, there were disputes between various versions of set theory and several technical problems concerning these theories, but it was felt that generally these theories enlightened the nature of sets to a satisfactory extent.[5]

The most basic axiom of every set theory states that sets are uniquely determined by their elements – that sets with the same elements are identical. Then there are axioms stipulating the existence of sets: of an empty set, of the set of all subsets of a given set, a set of all elements of all subsets of given set, etc. Then there are some more complicated axioms, the need for which you will not understand unless you seriously submerge yourself into set theory. These are not important for us here.

What is important for us is that plus/minus all entities that are addressed by modern mathematics can be grasped as various sets. (We will discuss, in Section 5.5, the claim of the logician Pavel Tichý that he needs a kind of entity that cannot be accommodated within set theory, but this claim is controversial.) Crucial for this is the appropriation of the concept of function, which is central for many areas of mathematics.

## 1.4 Frege's maneuver

Frege, however, did not believe that all meanings are objects. Objects are meanings of *names* (in the broadest sense of the word – even sentences, by Frege's lights, are sorts of names, meaning peculiar objects – truth values), but predicates express *concepts*, which, according to Frege, are certain functions; and functions, Frege

---

[5] See Lavine (1994); Grattan-Guinness (2000); Potter (2004).

maintains, are *not* objects. The thing, however, is that subsequent to this functions come to be ever more identified with what Frege called their courses-of-values and which *are* certain sets. Hence, even in this way Frege paved the way to grasping meanings as set-theoretical objects. And it is extremely important to understand how functions got into the picture.

Concepts, according to Frege, are functions which map objects on truth values. Think about how we use a predicate phrase: we attach to it a subject phrase to produce a sentence. Thus, we may take the phrase *conquered Gaul* and attach it to *Caesar* to get the (true) sentence *Caesar conquered Gaul* (Frege's famous example). Similarly, I can do it with other subject phrases:

*Caesar + conquered Gaul = Caesar conquered Gaul.*

*Aristotle + conquered Gaul = Aristotle conquered Gaul.*

*Cartman + conquered Gaul = Cartman conquered Gaul.*

...

We can summarize the behavior of the predicate as the following function:

*Caesar → Caesar conquered Gaul.*

*Aristotle → Aristotle conquered Gaul.*

*Cartman → Cartman conquered Gaul.*

...

All this concerns expressions; but we can shift this whole consideration from their level to the level of meanings (using $\|...\|$ to designate a meaning):

$\|Caesar\| \rightarrow \|Caesar\ conquered\ Gaul\|$

$\|Aristotle\| \rightarrow \|Aristotle\ conquered\ Gaul\|$

$\|Cartman\| \rightarrow \|Cartman\ conquered\ Gaul\|$

...

Now, in view of the fact that the meaning of a name, according to Frege, is the individual named by it, and the meaning of a sentence is its truth value, this boils down to (let me use pictures of individuals instead of names to stress that these are *not linguistic* objects)

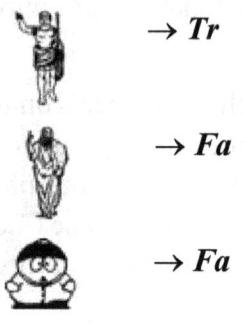

    → *Tr*

    → *Fa*

    → *Fa*

...

I propose calling this encapsulation of the functioning of an expression into a function *Frege's maneuver*. It has been copiously repeated by the formal semanticists in the building of their semantic models of natural language. Consider, for example, the adverb *quickly*, which maps predicative phrases on predicative phrases:

*conquered Gaul* → *conquered Gaul quickly*

*killed Cesar* → *killed Cesar quickly*

*robbed a bank* → *robbed a bank quickly*

...

By Frege's maneuver we can transform its semantic behavior into a function that maps functions from objects to truth values on the same kinds of functions. And in a similar way we can accommodate expressions of plenty of other categories within the semantics.

Something close to Frege's maneuver also played a crucial role in the transfer from the extensional to the intensional model of meaning. The idea behind this was that to understand an expression, I need to know not only its actual extension but also its potential extension in contrafactual situations. Thus, to know the extension of a sentence is to know its truth value, to know its intension is to know in which circumstances it is true (*viz.* to know its truth conditions).

Now the maneuver here close to Frege's encapsulates this again into a function – a function whose arguments are no longer meanings of linguistic expressions, but rather possible worlds. (What exactly the possible worlds are supposed to be has been a notorious source of dispute, which makes the foundations of this very approach somewhat shaky.) We will discuss this in detail in Chapter 4.

## 1.5 Semantic models of language

If meanings are ideal objects, hence if they neither exist within space and time nor exist merely in the subjective mental world of a speaker, they are difficult to get hold of and it is difficult to explain their workings. Frege, in effect, proposed to embed these objects into the realm of mathematics, which was later established as a realm supervised by set theory. This has turned out to be a fruitful path – but has this movement resolved the general questions concerning the nature of meanings in the sense that meanings *are* sets? We will leave the answer to this general question to the last chapter of the book; at this point, we will only conclude that it has turned out to be useful to *reconstruct* meanings as sets.

The notion of reconstruction lets us maintain a certain distance between the possibly "ineffable" meaning as such and its set-theoretic reconstruction. Therefore, I think it is best to see the situation where we reconstruct meanings as set-theoretical objects as building logico-mathematical *models* of language, especially of its semantics.

Let me, before I characterize the kind of models we are going to build, point out some of their general, structural features. We want there to be one and only one meaning for every meaningful expression (this is obviously an oversimplification, we are disregarding ambiguity); moreover, we want that the meaning of a complex expression be produced out of the meanings of its parts. Hence our models will incorporate what is called the *principle of compositionality*.[6] This principle states that the meaning of a complex expression is always uniquely determined by the meanings of its parts plus the mode of their combination. In particular, it states that if we denote the meaning of $E$ as $\|E\|$, then for every syntactic rule $R$ there must exist an operation $R^*$ such that

$$\|R(E_1,...,E_n)\| = R^*(\|E_1\|,...,\|E_n\|)$$

for every expressions $E_1, ..., E_n$ that can be combined by the rule R into a complex expression. The principle of compositionality is thus a constitutive feature of our semantic models. The principle of compositionality is equivalent to what can be called the principle of intersubstitutivity of synonyms:

if $\|E_i\| = \|E_i'\|$, then $\|R(E_1,...,E_i,...,E_n)\| = \|R(E_1,...,E_i',...,E_n)\|$.

Indeed, if the principle of compositionality holds and $\|E_i\| = \|E_i'\|$, then

$$\|R(E_1,...,E_i,...,E_n)\| =$$
$$R^*(\|E_1\|,...,\|E_i\|,...,\|E_n\|) =$$
$$R^*(\|E_1\|,...,\|E_i'\|,...,\|E_n\|) =$$
$$\|R(E_1,...,E_i',...,E_n)\|.$$

Conversely, the $R^*$ claimed by the principle of compositionality does not exist iff there are $E_1,...,E_n,E_1',...,E_n'$ so that $\|E_1\| = \|E_1'\|$, ..., $\|E_n\| = \|E_n'\|$ and $\|R(E_1,...,E_n)\| \neq \|R(E_1',...,E_n')\|$; but it is easy to see that this is excluded by the principle of the intersubstitutivity of synonyms.

As Janssen (1986) pointed out, given the principle of compositionality, we can see a semantically interpreted language as a many-sorted algebra of expressions mapped – by a homomorphic mapping – onto the many sorted algebra of meanings.

---

[6] See Werning, Machery, and Schurz (2005). See also Peregrin (2001a, Chapter 4) for a discussion of the motivations and consequences of the principle.

Speaking about such general structural semantic principles, we can mention one more, interconnecting meaning and truth. It states that if two sentences differ in truth value, they cannot but differ in meaning. If we denote the truth value of the sentence $S$ as $|S|$, we have[7]

if $|S_1| \neq |S_2|$, then $\|S_1\| \neq \|S_2\|$,

for every sentences $S_1$ and $S_2$.

This principle together with the principle of the intersubstitutivity of synonyms yields what can be called the principle of the intersubstitutivity of synonyms *salva verirate*:

if $\|E_i\| = \|E_i'\|$, then $|R(E_1,...,E_i,...,E_n)| = |R(E_1,...,E_i',...,E_n)|$,

whenever $R(E_1,...,E_i,...,E_n)$ and $R(E_1,...,E_i',...,E_n)$ are sentences. The inversion of this principle is sometimes called the Leibniz principle:

if $|R(E_1,...,E_i,...,E_n)| = |R(E_1,...,E_i',...,E_n)|$ for every sentences $R(E_1,...,E_i,...,E_n)$ and $R(E_1,...,E_i',...,E_n)$, then $\|E_i\| = \|E_i'\|$.

## 1.6 The four kinds of models

Let me explain, in greater detail, what kinds of models of formal semantics have been introduced and that we are going to reconstruct. Each model will consist of two parts: *syntax* and *semantics*. The syntactical part will reconstruct the set of well-formed expressions with which it will work. It too will consist of two parts: a vocabulary and a set of syntactic rules. The vocabulary will be a list of words, of elementary expressions divided into syntactic categories (e.g. the words Eco and Schwarzenegger of the category *term* and the words writer and actor of the category *predicate*). The rules will prescribe how to build more complex expressions out of simpler ones (e.g. that it is possible to

---

[7] Elsewhere, I called it *the principle of verifoundation* (see Peregrin 1994; 2001b). Cresswell (1982) considers it to be the most certain principle of semantics.

combine a term with a predicate into a *statement*: <u>writer</u>(<u>Eco</u>)). In sum, the syntactical part will determine what counts as a well-formed expression.

The semantical part of the definition of the model will equip every well-formed expression with a set-theoretical object, reconstructing the meaning of the expression (e.g. the words <u>Eco</u> and <u>Schwarzenegger</u> with the persons Eco resp. Schwarzenegger, and the words <u>writer</u> and <u>actor</u> with the sets of all writers resp. actors). It will consist of two parts copying the two parts of the syntactic component. The first part will assign a set-theoretical object to every word of the vocabulary. The second part, then, will tell us, for every syntactic rule, how to construct an object assigned to a complex expression built according to this rule from the objects assigned to its parts (e.g. that a statement consisting of a term and a predicate is assigned the truth value ***Tr*** if the object assigned to its term is an element of the set assigned to its predicate; and the value ***Fa*** otherwise).

What we call models are thus kinds of artificial languages such as those we know from formal logic. And indeed, I think that the artificial languages of logic can be seen as such kinds of models of natural language (Peregrin, 2020). The difference is that logical languages concentrate on a logical part of their vocabulary, usually leaving the extralogical part completely aside; whereas here we are interested in all parts of the vocabulary, without a difference.

In this book, I discuss four kinds of models delivered by formal semantics. In Chapter 3, I discuss what I call the *extensional model of meaning*. This is the model which was mostly developed within formal logic thanks to people like Frege, Tarski and many others. It is the model that is sufficient if we want to discuss mathematics or just leave aside the empirical dimension of language. Many logicians did concentrate on mathematics, and so they were content with this kind of semantics; however, for philosophers and linguists who took natural language seriously this kind of semantics was simply a non-starter.

The model that I call *intensional*, then, was a breakthrough. It was foreshadowed in the writings of Carnap and it came to full fruition in the hands of Montague and his followers. It showed how to model the meanings of even the empirical expressions, using the concept of possible world. This concept was hinted at by Carnap, who felt that extension is not a fair model of meaning in the intuitive sense of the word and that we need to get a grip on the notion of intension; independently of this, it was also arrived at by Kripke, who looked for the semantics of modal logic. I discuss this semantic model in Chapter 4.

As the intentional logical model turned out to not be wholly waterproof (it was challenged especially with the propositional attitude reports), there appeared amendments, and these led, quite quickly, to models which I classify as *hyperintensional*. This is a bundle of different kinds of models that share the assumption that meaning has a kind of structure that is related to the syntactic structure of their expressions. These models are discussed in Chapter 5.

Finally, in Chapter 6, I discuss the models that I call *dynamic*. They appeared most recently, and they took into consideration the fact that language is a means of discourse which is a dynamic enterprise. Their proponents maintained that if we want to analyze natural language with all its peculiarities (such as pronouns or the phenomenon of anaphora), we must build different semantic models than the intensional or the hyperintensional ones. In particular, we need to incorporate the concept of context.

In the last chapter of the book, I return to the general problem of capturing meaning as an object. I conclude that the question whether this is reasonable must be kept apart from the question whether meaning really *is* an object (where there may be no clear answer to the latter question). I maintain that we should see the relation between the models of meaning and their target phenomena as that of *explication*. And that this explication is extremely fruitful because it equips semantics with a huge toolbox of set theory.

# 2 A very short history of formal semantics

## 2.1 Frege

The roots of formal semantics, we already saw, can be traced back to the writings of Gottlob Frege (1848-1925), the German mathematician, logician and philosopher who laid the foundation not only of modern formal logic but also of what has later become known as analytic philosophy (Dummett, 1996). He was the first to clearly realize that semantics has little to do with psychology, and that it could be usefully explicated in mathematical terms (Dummett, 1981a; 1981b).

Frege's depsychologization of semantics followed from his depsychologization of logic. Frege understood how crucial it was for the development of logic to draw a sharp boundary separating it from psychology: to make it clear that logic is *not* a matter of what is going on in some person's head in the sense that psychology is. The reason is that logic is concerned with what is true and consequently what follows from what – and whether something is true, or whether something follows from something else, is an objective matter independent of what is going on in the head of a particular person.

As a consequence, Frege realized that if logic must be separated from psychology then the same is true for semantics – at least insofar as semantics underlies truth and entailment. It is clear that the truth value of a sentence depends on the meaning of the sentence: the sentence "London is in England" is true not only due to the fact that London is in England, but also of course because the words that it consists of mean what they do in English. Hence, if meaning were a matter of what is going on in somebody's head, then truth would also have to be – hence meaning must not, on pain of the subjectivization of truth, be a psychological matter. But what, then, *is* meaning?

Frege started from the *prima facie* obvious fact that names stand for objects of the world. (Unprecedentedly, he assimilated sentences to stand for this as well: he saw them as specific kinds of names that denote truth values. The reason for this move was that he divided expressions into two sharply separated groups: into "saturated" – i.e. self-standing – and "unsaturated" – i.e. incomplete – ones. He took names and sentences as species of the former kind, whereas he took predicates as paradigmatic examples of the latter one; and he came to use the word "name" as a synonym of "saturated expression".) His most brilliant contribution to the explication of the concept of meaning then was the way he accounted for the meanings of predicates in terms of what we have dubbed Frege's maneuver. He called them *concepts*, per usual; but he rejected the usual way of seeing concepts as something mental and, in effect, he suggested explicating them by means of studying how the expressions which express them – i.e. predicates – function within language.

What is the function of a predicate, such as "to think"? Well, the predicate is attached to a subject to form a sentence. Hence, if we assume that the meaning of a complex expression is the result of combining the meanings of its parts (i.e. that meanings are composed in a way paralleling that in which the expressions expressing them are), then the meaning of the predicate together with the meaning of a subject, which is the object stood for by the subject, yields the meaning of a sentence, i.e. a truth value. A concept is thus something that together with an object yields a truth value – and this led Frege to identify concepts with functions, in the mathematical sense of the word, taking objects to truth values. In effect, this meant the identification of the meaning of an item with the semantic function of the item captured as a function in the mathematical sense of the word; and this opened the door for a mathematical treatment of semantics. Thus, we can say that Frege married semantics, which he had earlier divorced from psychology, to mathematics.

Notice that Frege's maneuver has two substantial presuppositions. There is the presupposition that the meaning of a complex expression is yielded by (or "composed of") meanings of the parts of the expression. This is the *principle of compositionality* we discussed in §1.5. How do we know that this principle holds? Some theoreticians appear to think that it is an empirical thesis that must be verified as empirical theses are: by means of inspecting as many cases as possible. However, such a view presupposes that meanings are independently identifiable objects whose combinations can be studied in the way we study, e.g. combinations of molecules in a solution. That presupposes that we can empirically verify (or falsify) the thesis that, say, the meaning of a sentence is yielded by the meaning of its subject and that of its predicate by means of finding the meanings and finding out what they yield if they are put together. In contrast to this, we saw that for Frege the principle was rather a way of articulating what it takes to be meaning: the principle was co-constitutive of the notion of meaning in the sense analogous to that in which, say, the principle of extensionality is co-constitutive of the concept of set. And just as it makes no sense to try to find out whether sets are extensional (for this is simply part of what it takes to be a set), it makes no sense to try to find whether meaning is compositional.

The other presupposition of Frege's maneuver is of a different kind: it concerns the behavior of the particular expressions to which it is applied. The presupposition is that the role of the expression within language is exhausted by, or at least in some sense reducible to, its role within the kind of syntactic combination which is taken into consideration. We explicated the meanings of predicates by considering the way they combine with names into sentences; but predicates also do other things, e.g. combine with adverbials into complex predicates. We must always be sure that this is taken care of; that it is proven that it is somehow substantiated to treat some part of the functioning of an expression as representative of the whole functioning.

Is Frege's way of explicating the concept of meaning acceptable? In fact, it is not: what Frege called *meaning* cannot be taken as a plausible explication of the pre-theoretic notion of meaning. After all, who would want to claim that all true sentences have the same *meaning*? And Frege soon came to realize the implausibility of such an explication. He therefore complemented his theory of meaning by what he called a theory of *sense*.⁸ Every name, he claimed, has not only a meaning, but rather also a sense, which is the "way of givenness" of the meaning. And it is therefore Frege's concept of *sense*, rather than his concept of *meaning*, which is to be taken as his explication of the intuitive concept of meaning.

Frege's own instructive example is that of the terms "morning star" and "evening star". As we now know, these two terms refer to one and the same celestial body, the planet Venus. Hence, in Frege's sense of the word they share the same meaning or, in the current jargon, the same *referent*. However, although the sentence "The morning star is the morning star" is obviously trivial, "The morning star is the evening star" does not appear to be such. The reason, Frege claimed, is that the terms differ in their senses, i.e. in the ways they present their referent: "the morning star" presents it as the most attractive body in the morning sky, whereas "the evening star" presents it as the most attractive one in the evening sky.

Hence we have the general picture according to which the relation between a name and what the name refers to is mediated by the sense of the name:

$$\text{NAME}$$
$$\downarrow$$
$$\text{SENSE (i.e. meaning in the intuitive sense of the word)}$$
$$\downarrow$$
$$\text{MEANING (i.e. object referred to)}$$

---

[8] See Frege (1892b).

Another of Frege's path-breaking contributions to the development of logic and mathematics was his establishment of a logical language which, even though it looked very different, was structurally almost identical to what we now call predicate calculus. His basic achievement was the introduction of what we know today as quantifiers (although Frege's own notation was idiosyncratic).[9]

When introducing quantifiers, Frege said roughly this: Imagine a sentence decomposed into two parts, and imagine one of the parts "abstracted away", the sentence being thereby turned into an "unsaturated", gappy torso. Then imagine the gap in this matrix being filled with various things[10] and consider the truth values resulting from the individual saturations. In some cases, it can happen that however we fill the gap, we will always reach a true sentence; and this Frege abbreviated by means of the general quantifier. Thus, the sentence that in modern notation reads

$\forall x F x$

was, by definition, his shorthand for "whatever replaces the $x$ in F$x$, we gain a true sentence". Similarly, Frege's equivalent of the modern

$\exists x F x$

was introduced to shorten the claim that there is at least one thing that can replace the $x$ in F$x$ so that we gain a true sentence.

In this way, Frege introduced the machinery of quantifiers and variables in the shape that has underlain almost all formal logic since.

---

[9] See Frege (1879).

[10] According to modern standards, this is ambiguous: it can mean either (i) imagine that the gap is literally filled with an *expression*, or (ii) imagine that the formal sign marking the gap, the variable, is made to refer to an *object*. These two interpretations (ultimately resulting in what is nowadays called the *substitutional* and the *objectual* notion of quantification, respectively) come out as equivalent only if we assume that every object (within the relevant universe of discourse) has a name.

## 2.2 Tarski

Frege's distinction between *meaning* and *sense* has influenced virtually all subsequent theories of meaning; however, his "mathematical" explication of concepts and meanings has not been absorbed as quickly as it might have deserved. It is thus somewhat peculiar that when Carnap (who was familiar with Frege's teaching) in 1934 wrote his important book about the logical formalization of language, *Der Logische Syntax der Sprache*, he claimed that the only aspect of language that is susceptible to formalization is syntax; that semantics is ineffable. It was only slightly later, after he had absorbed the teaching of Alfred Tarski (1901-1983), that he admitted that there was a way of formalizing semantics that was as rigorous as the formalization of syntax.

A great deal of Tarski's theory of semantics looks like re-discovering Frege's ideas and putting them into the context of a more developed theory of formal logic. His formalization of semantics was a kind of by-product of his theory of truth. What he was after was the fixation of the meaning of "true" by putting together some *axioms* governing it; just like some of his colleagues had fixed the meaning of "set" by means of axiomatic set theories before. Tarski (1935; 1944). realized that what would fix the meaning of "true" were all sentences of the shape

(1) **true**(....) $\leftrightarrow$ ___,

with the dots replaced by a name of a sentence and the underscore by the very sentence. Sentences of this form are now generally called T-sentences (you may opt for interpreting the "T-" as standing either for "Tarski" or for "truth"). However, the set of all these sentences was infinite, and hence could not be taken as the desired theory. The problem of the explication of the concept of truth appeared then to boil down to the problem of finding a "reasonable" (preferably finite) set of axioms that would entail the infinite set of all the T-sentences.[11]

If we assume that the language which we are considering is the language of standard logic, we can divide its sentences into three classes: *atomic* (sentences consisting of a predicate applied to the appropriate number of terms), *logically complex* (those consisting of a logical operator applied to one or two sub-sentences) and *quantified* (those consisting of a quantifier binding a variable in a formula). Now it is clear that, using a few simple principles, we can deduce the T-sentences for logically complex sentences from those for the other sentences. The point is that using the principles (where $Neg(x)$ refers to

---

[11] Could we make do with a single axiom produced by "closing" the schema (1) by means of variables and quantifiers? It is easy to see that $\forall x \forall y(\mathbf{true}(x) \leftrightarrow y)$ would not do; but what about $\forall y(\mathbf{true}('y') \leftrightarrow y)$? A moment's reflection reveals that it would work only if '$y$' referred to a name of the sentence $y$ – but in fact the name which arises from enclosing a symbol in quotes notoriously refers to the very symbol. (Hence '$y$' does not refer to the name of the sentence $y$, but rather to the penultimate letter of the alphabet.) Similarly $\forall y(\mathbf{true}(\mathbf{name}(y)) \leftrightarrow y)$: even if we disregard that it would require $y$ to be a sentential variable and **name** to form terms out of sentences (and hence would not be accommodable within the framework of the predicate calculus, which Tarski took as the ultimate framework of logic), there does not appear to be a *function* taking denotations of sentences to their names (which is to be denoted by **name**) – the relationship between the former and the latter appears to be one-to-many. And even if we assumed that for every denotation of sentence there is one "canonical" name to be yielded by such a function, to make the functor denote the right function (taking a denotation of a sentence to its canonical name) would amount to inventing an axiomatic theory doing precisely what Tarski urged: entailing all the T-sentences. This becomes obvious when trying to engage the converse functor **denotation** forming sentences out of terms and denoting a function taking sentences to what they denote. In this case the T-scheme would yield us $\forall y(\mathbf{true}(y) \leftrightarrow \mathbf{denotation}(y))$; and it is clear that then **denotation** becomes simply *equivalent* to **true**, and providing a theory for it *is* providing a theory of **true**. See also Kirkham (1995, Chapter 5).

the negation of the sentence $x$ and $Con(x,y)$ refers to the conjunction of $x$ and $y$)[12]

(2) **true**($Neg(x)$) ↔ not **true**($x$),

and

(3) **true**($Con(x,y)$) ↔ (**true**($x$) and **true**($y$)),

we can deduce the T-sentence of any conjunction and any negation from those of its immediate subsentence(s) (the same clearly holds for the other standard logical operators); and applying this recursively, we can eventually deduce it from other than logically complex sentences.

Now what about quantified sentences of the kind $\forall xF$? In general, they do not contain any subsentences but only (possibly open) subformulas, so the previous method is not applicable. Here is where Tarski came in with an ingenious idea[13]: he put the concept of truth to the side for a while and instead turned to the notion of *satisfaction*, a relation between formulas and objects of the world. Intuitively, satisfaction is the relation which holds between the formula **brothers**($x,y$) and a pair of persons iff the two persons are brothers; but as a formula can contain an unlimited number of different variables, things are simplified if we define it as a relation between formulas and infinite sequences of objects. (Variables are thought of as linearly – e.g. alphabetically – ordered and hence corresponding to the objects of the sequences in a one-one way. Hence if we assume that $x$ is the first and $y$ the second variable, in the case of **brothers**($x,y$) only the first two elements of such a sequence matter.)

---

[12] Note that talking about the sentences of a language requires that the sentences are contained within our universe of discourse. As the assumption that we are capable of treating a language within the very language might be dangerous (by being liable, as Tarski showed, to inducing inconsistencies), we assume that we are operating within is what Tarski called *metalanguage*, i.e. a language capable of treating of an *object language* by containing the names of its expressions.

[13] See Peregrin (1999).

It is easy to see that satisfaction of quantified formulas is reducible to that of their unquantified subformulas. To show this, we must first introduce some notation. Let us assume that the variables of the language we deal with are ordered and let Var($i$) refer to the $i$th one. Moreover, let All($x,y$) denote the formula constituted by the concatenation of the general quantifier, a variable $x$ and a formula $y$; and let Ex($x,y$) denote the formula constituted by the concatenation of the existential quantifier, the variable $x$ and the formula $y$.[14] Finally, if $S$ is an infinite sequence of objects, then let S[$i,a$] refer to the sequence which is identical with $S$ save for the only possible difference that its $i$th constituent is $a$. Then it clearly holds that

(4) **sat**(All(Var($i$),$y$),$S$) ↔ **sat**($y$,S[$i,a$]) for every object $a$ of the universe; and

(5) **sat**(Ex(Var($i$),$y$),$S$) ↔ **sat**($y$,S[$i,a$]) for at least one object $a$ of the universe

and hence satisfaction for quantified formulas is indeed reducible to that for their subformulas. (As it is more perspicuous to work directly with functions assigning objects to variables instead of the object-sequences which effect such assignments indirectly, we will do so; and we will call a function assigning an object to every variable a *valuation of variables*.)

---

[14] Note that the sign "$x$", as employed within the previous paragraph, is not a variable (of the object language) but a *name* of a variable (a symbol of our metalanguage). This may easily lead to a certain chaos, for the following reason: When we speak about a non-linguistic object, we have no choice but to use a sign standing for it – we cannot put, e.g. an apple itself into a sentence speaking about it. Not so, however, in case of *linguistic* objects, signs. We can, as is often done, put the object itself, instead of its name, into a sentence. Thus, suppose that "α" is a variable of the language we are investigating. Then we can say *the formula ... contains the variable* "α", or, if "α" is the first variable according to the relevant ordering and we use the notation introduced in the previous paragraph, we can equivalently say *the formula ... contains the variable Var*(1), but it is often also written *the formula ... contains the variable α*. The last formulation is literally incorrect, but it is often used instead of the first one.

Moreover, satisfaction for logically complex formulas is reducible to that for logically simple ones along the lines wholly analogous to those along which truth is, *viz.*

(6) $\mathbf{sat}(Neg(x),S) \leftrightarrow$ not $\mathbf{sat}(x,S)$

(7) $\mathbf{sat}(Con(x,y),S) \leftrightarrow (\mathbf{sat}(x,S)$ and $\mathbf{sat}(x,S))$.

This implies that for a language with a finite number of atomic formulas we can have a *finite* theory of satisfaction (the theory would be constituted by the sentences stating the satisfaction conditions for all the atomic formulas plus (4)-(7)). Now the point of this maneuver is that for sentences (formulas with no free variables), truth is clearly reducible to satisfaction:

(8) $\mathbf{true}(x) \leftrightarrow (\mathbf{sat}(x,S)$ for every sequence $S)$.

Hence the finite theory of satisfaction yields us the desired finite theory of *truth*.

Tarski's investigation thus seemed to suggest that the concept of truth must be attacked by means of the investigation of a language-world relation such as satisfaction – therefore this theory has come to be called the *semantic* theory of truth.[15] Moreover, for many logicians (notably for Carnap) it acted as a revelation of the fact that semantics was not as inaccessible to a formal treatment as it had appeared up to that point.

## 2.3 Carnap

Tarski's relation of satisfaction gestures towards a formalization of the relation expression-meaning (or expression-referent), but it is not really a formalization of it. In fact, from the viewpoint of natural language it is slightly unnatural – for it presupposes the existence of open formulas that have no counterpart in natural language.[16]

---

[15] See Kirkham (1995, Chapter 5); Peregrin (1999).

[16] Cf. Peregrin (2000b).

Let us consider a formal language with no variables and quantifiers, but with an infinite number of atomic sentences. Let us assume that the category of terms of the language is productive, i.e. that we have functors capable of taking terms to terms (such as '+' or '×' of arithmetic, which join pairs of terms into complex terms). How might a truth definition for such a language look? Instead of Tarski's **sat** we would need the relation **des** relating an expression to an object, to what it "designates". Let us provide for the reducibility of **true** to **des** (which gives the introduction of **des** its point) by assuming that the objects that are designated by sentences, "propositions", are capable of "being true". As the analogue of T-sentences we now have what can be called D-sentences, namely sentences of the form

**des**(...., __),

where the dots are replaced by the name of an expression and the underscore by the expression itself. Thus, the D-sentences include

**des**("John", John)

**des**("to be bald", to be bald)

**des**("John is bald", John is bald)

the second arguments of which are supposed to stand for the individual John, the property of being bald and the proposition that John is bald (whatever properties and propositions might be supposed to be here), respectively. In this way we reach the view of semantics developed in the 1940s by Rudolf Carnap (1891-1970).

Now if the concept of designation is to yield us a theory of truth as the concept of satisfaction did, we must provide for two things: (1) a finite theory of designation (a finite number of axioms entailing all the D-sentences); and (2) the reduction of the concept of truth to the concept of designation. Let us start with the latter.

Carnap (1942, p. 51) claims that the reduction works as follows[17]:

---

[17] Note that if we define **designation**($x$) as the only *prop* such that **des**($x$,*prop*), this becomes tantamount to the last proposal discussed above in footnote 12.

**true**(*x*) ↔ ∃*prop* (**des**(*x*,*prop*) ∧ *prop*)

i.e. a sentence *x* is true iff it expresses a proposition *prop* and *prop* is the case. Hence, a sentence is true if, e.g., it expresses the proposition that it is raining and it is the case that it is raining. This amounts to assuming that there is a way from propositions to their truth values. The trouble is that it is not clear what propositions (and properties) are really supposed to be. Alternatively, we could stick to the Fregean approach and to assume that what sentences designate are directly the truth values, which would yield us

**true**(*x*) ↔ **des**(*x*,*Tr*).

Then the reduction of all the D-sentences to the D-sentences for logically non-complex sentences is effected by the principles of the following kind

**des**(*Neg*(*x*),*Tr*) ↔ **des**(*x*,*Fa*)

**des**(*Con*(*x*,*y*),*Tr*) ↔ (**des**(*x*,*Tr*) and **des**(*x*,*Tr*)).

Now, however, we need not stop here, for designation is defined not only for sentences, but also for their parts; and for the sentence *p*(*s*) consisting of a subject *s* a predicate *p* we can stipulate

**des**(*p*(*s*),*Tr*) ↔ ∃*i*∃*r*(**des**(*s*,*i*) and **des**(*p*,*r*) and the individual *i* has the property *r*).

If we accept the Fregean identification of properties with functions, we can turn this further into

**des**(*p*(*s*),*Tr*) ↔ ∃*i*∃*r*(**des**(*s*,*i*) and **des**(*p*,*r*) and *r*(*i*)).

Moreover, for complex names of the shape *f*(*t*) we have

**des**(*f*(*t*),*i*) ↔ ∃*g*∃*i*′(**des**(*t*,*i*′) and **des**(*f*,*g*) and *i* = *g*(*i*′)).

Carnap (1947) also realized that if what we are after is meaning in the intuitive sense of the word, then we should not be interested so much in meanings in the sense of Frege, but rather in Fregean senses. However, as Frege did not explicate the concept of sense to Carnap's satisfaction, Carnap proposed replacing the Fregean twin concepts of meaning and sense with the concepts of *extension* and *intension*. Roughly speaking, the *extension* of a term is what the term shares with all terms that are equivalent to it; whereas its intension is what it shares with all the terms that are *logically* equivalent to it.

Of course, this definition becomes non-trivial only after we give a rigorous account of the concept of equivalence on which it rests. For the basic categories of the predicate calculus this is not difficult: two individual expressions $I_1$ and $I_2$ are equivalent iff $I_1 = I_2$, two $n$-ary predicates $P_1$ and $P_2$ are equivalent iff $\forall x_1...\forall x_n(P_1(x_1,...,x_n) \leftrightarrow P_2(x_1,...,x_n))$, and two sentences $S_1$ and $S_2$ are equivalent iff $S_1 \leftrightarrow S_2$. This explication leads to a concept of extension almost indistinguishable from Frege's concept of meaning: the extension of an individual expression being the object for which it stands, that of a predicate the function assigning the truth value **Tr** to those $n$-tuples of objects of which the predicate is true (or, equivalently, the class of the $n$-tuples) and that of a sentence its truth value.

The concept of intension is far more problematic, but Carnap indicated a way to approach it, namely via the concept of (possible) *state-of-affairs*. This concept has later been replaced, especially thanks to the seminal results of Saul Kripke (1963b) concerning the model theory for modal propositional calculus, by the concept of *possible world*.

## 2.4 Standard Logic and its Semantics

Some logicians during the twentieth century, especially those inclined toward mathematics, came to the conclusion that there is something like *the* language of logic – that it is the language of what has come to be called first-order predicate calculus.[18] The vocabulary of a language within the framework of this calculus falls into three categories:

---

[18] Probably not many logicians would subscribe to the existence of "one true logic" explicitly (as e.g. Priest, 2001, would), but in the majority of contexts where the term "logic" is used without a qualification, it means first-order

1. logical constants ($\neg, \wedge, \vee, \rightarrow, \exists, \forall$);
2. extralogical constants (individual, predicate, functor);
3. variables (individual).

The syntax of such a language is then as follows:

An individual constant and an individual variable is a *term*; moreover, if $F$ is an $n$-ary functor and $T_1,...,T_n$ are terms, then $F(T_1,...,T_n)$ is a term (and nothing else is a term).

If $P$ is an $n$-ary predicate and $T_1,...,T_n$ are terms, then $P(T_1,...,T_n)$ is a *formula*; moreover, if $F_1$ and $F_2$ are formulas and $x$ is a variable, then $\neg F_1$, $F_1 \wedge F_2$, $F_1 \vee F_2$, $F_1 \rightarrow F_2$, $\forall x F_1$ and $\exists x F_1$ are also formulas (and nothing else is a formula).

The semantic treatment of a language of this kind which has become standard needs some elucidation, for it rests heavily on the logical/extralogical boundary, which has not yet played a principal role in our exposition. We have seen that Tarski's semantic theory of truth was based on the concept of satisfaction, which in turn necessitated the employment of the valuations of variables. Tarski's explication of the concept of logical consequence was based on an analogous idea, only applied at a higher level. While a sentence $S$ is a consequence (*simpliciter*) of $S_1,...,S_n$ iff $S$ cannot be false unless at least one of $S_1,...,S_n$ is, it is their *logical* consequence iff this is the case independently of what is stood for by the non-logical words in $S_1,...,S_n$ and $S$ – in other words, if this is the case for every assignment of values to these words. (This, of course, presupposes that we can classify the vocabulary of our language into the logical and the non-logical part, which is far from non-controversial, but let us ignore this problem for now.) This means that the fact that *John is unmarried* is a logical consequence of *John is a bachelor* and *Every bachelor is unmarried* because *X is (a) Y* is a consequence of *X is (a) Z* and *Every Z is (a) Y* whatever may be stood for by *X*, *Y* and *Z*, schematically

---

predicate calculus.

$\forall X \forall Y \forall Z$ (*X is (a) Z, Every Z is (a) Y* $\Rightarrow$ *X is (a) Y*).

Now this is a "meta-level" quantification (the "quasiformula" just presented is not to be understood as belonging to the language in question, but rather as being "about" it) that cannot be mixed with the "object level" one (such as would result, e.g., from the standard regimentation of *Every bachelor is unmarried* as $\forall x(B(x) \rightarrow U(x))$). Therefore, we need *two* separate kinds of variable words and *two* separate sets of valuations: we keep calling the "object level" variables simply *variables*, whereas we call the "meta-level" ones *parameters*. While the former underlie quantification of the object language, the latter underlie (explicit or implicit) quantification of the metalanguage, of statements *about* the object language.

Thus, the resulting semantics is based on two sets of assignments of objects to expressions: the *valuation of variables* and the *interpretation of parameters*. This variety of formal semantics has been developed especially within the framework of what has come to be called *model theory* and which has developed out of Tarski's later work. Hence, from the model-theoretic perspective, a language of the kind discussed has three basic kinds of expressions: The semantics of logical constants is taken to be fixed, and they are usually not taken to designate objects (although it also is possible to take them so). Parameters or extralogical constants are taken to be assigned a denotation by the interpretation, which maps individual constants on elements of a universe, predicate constants on relations over the universe, and functor constants on functions over the universe. Variables are then taken to be interpreted by the valuation, which maps them on the elements of the universe. Given an interpretation $I$ and a valuation $V$, every individual term $T$ is assigned a denotation $\|T\|_{I,V}$ based on $I$ and $V$ in the following way:

If $T$ is an individual constant, then $\|T\|_{I,V} = I(T)$

If $T$ is an individual variable, then $\|T\|_{I,V} = V(T)$

If $T$ is $U(T_1,...,T_n)$ for some functor $U$ and terms $T_1,...,T_n$,
then $\|U(T_1,...,T_n)\|_{I,V} = I(U)(\|T_1\|_{I,V},...,\|T_n\|_{I,V})$.

An interpretation *I* and a valuation *V* also render each formula true or false. The usual inductive definition goes as follows:

If *F* is $P(T_1,...,T_n)$ for some predicate *P* and terms $T_1,...,T_n$, then *F* is true w.r.t. (or satisfied by) *I* and *V* iff $<\|T_1\|_{I,V},...,\|T_n\|_{I,V}> \in I(P)$

If *F* is ¬*F'*, then *F* is true w.r.t. (or satisfied by) *I* and *V* iff *F'* is not

If *F* is $F_1 \wedge F_2$, then *F* is true w.r.t. (or satisfied by) *I* and *V* iff both $F_1$ and $F_2$ are

If *F* is $F_1 \vee F_2$, then *F* is true w.r.t. (or satisfied by) *I* and *V* iff either $F_1$, or $F_2$ is

If *F* is $F_1 \rightarrow F_2$, then *F* is true w.r.t. (or satisfied by) *I* and *V* iff either $F_2$ is or $F_1$ is not

If *F* is ∀*xF'*, then *F* is true w.r.t. (or satisfied by) *I* and *V* iff *F'* is true w.r.t. (or satisfied by) *I* and *V'* for every valuation *V'* which differs from *V* at most in the value it assigns to *x*

If *F* is ∃*xF'*, then *F* is true w.r.t. (or satisfied by) *I* and *V* iff *F'* is true w.r.t. (or satisfied by) *I* and *V'* for some valuation *V'* which differs from *V* at most in the value it assigns to *x*.

The fact that this system of logic is sometimes accepted as the "standard", "classical" or "normal" logic should not hide the fact that there exist lots of both alternatives and of extensions. Especially in recent decades there have emerged an immense number of new logical systems; some of them having been initiated by impulses from formal semantics.

## 2.5 Chomsky

Both Tarski and Carnap saw an unbridgeable gap between natural language and the formal languages of logic they investigated. They claimed that natural languages, not being exactly defined, cannot be directly studied by the mathematical means developed by logicians; and they tacitly assumed that the formal languages they dealt with were what natural language should ideally be replaced by if we want to do serious science. Moreover, the later Tarski and his followers developing model theory were increasingly delving deeper into pure mathematics and losing sight of natural language.

However, at the same time and quite independently of the development of logic, a revolution within the approach to natural language, which was to result in a large scale "mathematization" of linguistics, was being started by Noam Chomsky (1928-). Chomsky's original goal was a rigorous description of the syntax of natural language. In his path-breaking 1957 work *Syntactic Structures*, he introduced a general framework for such a description. It is based on the concept of *generative grammar*, in effect a collection of rules understood as generating all well-formed sentences of the language being described.

The basic idea behind Chomsky's generative grammar is the idea of a *rewrite rule*. A rewrite rule simply instructs us to rewrite a sequence of symbols by another sequence of symbols. Thus, the rule

$S \rightarrow NP\ VP$

instructs us to rewrite "*S*" by "*NP VP*". Now the idea of a generative grammar for a language $L$ is the idea of a set of rewrite rules, working with the vocabulary of $L$ plus some set of auxiliary symbols, such that the set of all strings which can be produced by means of the (repeatable) application of the rules to the symbol $S$ and which contain no auxiliary symbols coincides with the set of all the well-formed sentences of $L$. Thus, if $L$ were to consist of the four sentences "John walks", "Mary walks", "John whistles", "Mary whistles", one of the possible generative grammars for it would be

$S \rightarrow N\ V$

$N \rightarrow$ John

$N \rightarrow$ Mary

$V \rightarrow$ walks

$V \rightarrow$ whistles

or, in an abbreviated form,

$S \rightarrow N\ V$

$N \rightarrow$ John | Mary

$V \rightarrow$ walks | whistles.

(Clearly, as long as the number of sentences of the language in question is finite there is always the trivial grammar consisting of the rules instructing us to rewrite $S$ by every particular sentence. However, as the number of sentences of natural language is potentially infinite, grammars for them cannot be that simple.)

Chomsky then supplemented rewrite rules by the so-called transformation rules, and subsequently introduced plenty of extensions, modifications and innovations of his model which refashioned its nature several times. But the basic idea remained unchanged: the formal grammar should provide for the generation of all and only well-formed sentences of the language under consideration.

What has changed is Chomsky's interpretation of the generative and transformative rules. At first they looked to be mere utensils of his theory that did not correspond to anything real, later they looked ever more like descriptions of something to be found in the human mind/brain, namely in that part of it which Chomsky called the *language faculty*.[19] Thus, what originally looked like a mostly abstract mathematics came to resemble an empirical theory of how language is realized in human brains.

Does Chomsky's approach go beyond what we know from the logical theories of formal languages? Not really. Consider the grammar of standard logic as summarized in the previous section. Its syntax can be, and usually is, defined in the following way:

An individual constant and a variable is a term.

If $F^N$ is an *n*-ary functor and $T_1,...,T_n$ are terms, then $F^N(T_1,...,T_n)$ is a term.

---

[19] See, e.g. Chomsky (1986; 1993; 2000).

If $P^N$ is an *n*-ary predicate and $T_1,...,T_n$ are terms, then $P^N(T_1,...,T_n)$ is a formula.

If $F_1$ and $F_2$ are formulas, then $\neg F_1$, $F_1 \wedge F_2$, $F_1 \vee F_2$, $F_1 \rightarrow F_2$, $\forall x F_1$, $\exists x F_1$ are formulas.

This yields us, rather straightforwardly, the following generative grammar:

$S \rightarrow P^N(T,...,T) \mid \neg S \mid S \wedge S \mid S \vee S \mid S \rightarrow S \mid \forall V S \mid \exists V S$

$T \rightarrow V \mid C \mid F^N(T,...,T)$

$P^N \rightarrow ...$

$F^N \rightarrow ...$

$V \rightarrow ...$

$C \rightarrow ...$

Hence, from this viewpoint the formalization of syntax proposed by Chomsky is only a minor variation on the theme of the specification of a formal language standardly entertained in logic. However, the path-breaking import of Chomsky's approach did not consist in the shape of his grammars, but in the fact that he wasn't proposing them to define formal languages but rather to account for natural ones; and that he managed to persuade a substantial part of the scientific community that the syntax of a natural languaged can be usefully captured by formal means.

Chomsky's successful attempt at the rigorization of natural language syntax was followed by attempts at the rigorization of semantics along analogous lines. Probably the first was the so-called *generative semantics* (Lakoff, 1971); Chomsky himself then extended his theory to cover not only syntax, but other "levels" of language as well. The idea behind such attempts was to capture "semantic structure" in a way analogous to the one in which the syntactic structure was captured. Many linguists did take this to be an acceptable approach to the semantics of natural language; however, there were also protests that theories of this kind do not amount to theories of semantic interpretation. The most substantial argument, leveled, e.g. by Lewis (1972), seemed to be that nothing can aspire to being a theory of *semantics* unless it yields a theory of truth conditions.

## 2.6 Montague and since

So, on the one hand, during a period centered around the 1960s there was a developed mathematical theory of natural language, but only of its syntax; and, on the other hand, there was a developed theory of semantics, but only for the standard predicate calculus, which appeared to be too simple to provide for an interesting model of natural language.

Of course there were logicians, linguists and philosophers who thought about bridging the gap. The most famous of them was the American logician Richard Montague (1930-1971), who proposed a logical system which, on the one hand, had a rigorously defined, Tarsko-Carnapian semantics, and, on the other, provided for a much more realistic model of natural language than any previous logical language. As a consequence, some theoreticians of language realized that model theory might therefore provide for an interesting explication of the semantics of natural language.

One important ingredient of this development was the construction, thanks to Kripke (1963a; 1963b; 1965), of a semantics for modal logics. Here is where the all-important concept of *possible world* appeared as a pillar of semantic theory. Modal logic is, roughly, the logic of necessity and possibility; and Kripke realized that to account for its semantics, we must not let the content of a sentence be exhausted by its truth-value (in the actual world), that it needs to also contain the information about its (potential) truth-values in worlds that are only possible, not actual.

Montague realized that if the model-theoretic means is to be engaged for the purpose of explication of the semantics of natural language, then it cannot stay on the level of extension. Drawing on the ideas of Carnap and Kripke, he formalized the concept of intension as, in effect, the relativization of extension to possible worlds (Montague, 1974). This is to say that he presupposed a given set of possible worlds (representing ways our world might also be) and understood an intension of an expression as the function taking a possible world to the extension of the expression within the world. Thus, if the extension of the singular term "the president of the USA" is the (current) president of the USA, its intension will be the function taking every possible world to the person who is the president of the USA in that world (if any); if the extension of the general term "horse" is the set of actual horses, its intension is the function which takes every possible world to the set of all the horses of the world; and if the extension of the sentence "The president of the USA is a horse" is the truth value $\boldsymbol{Fa}$, its intension is a function which takes every possible world to the truth value of the sentence in that world.

Montague furnished each expression of his model of language with a *denotation* (extension) and a *sense* (intension) and assumed that although what is essential are denotations, there are contexts in which the sense of an expression somehow assumes the place of its denotation. In particular, he introduced the operator $^\wedge$ such that for any expression $E$ the denotation of $^\wedge E$ is – by definition – the sense of $E$. (The dual operator $^\vee$ then worked the other way around: the sense of $^\vee E$ is defined as the denotation of $E$.) The idea was that a natural language expression E could, on the level of Montague's logic, be interpreted as either $E$ or $^\wedge E$, depending on its character and the context of its occurrence.

Moreover, Montague introduced a very general framework for the formalization of languages. We have seen that whereas standard model theory assumed a discriminating stance towards the vocabulary of its language (only words of some categories were taken as designators), the Chomskian theory of syntax took expressions of all categories alike. Montague assimilated model theory for his intensional logic to the indiscriminative stance. (This was not unprecedented: this stance was adopted long ago by people studying the semantics of lambda calculus, such as Church (1940) or Henkin (1949); on this, Montague's intensional logic also drew.)

This led Montague, in effect, to the vantage point from which an (uninterpreted) language appeared as a finitely generated algebra, its carrier being constituted by the well-formed expressions of the language, its generators being the words and its operations the syntactical, formation rules. The interpretation (meaning-assignment) for such a language then appears as a homomorphism of the algebra into another algebra (of "denotations" or "meanings"), the requirement of homomorphism reflecting the principle of compositionality (Janssen, 1986).

The Montagovian vantage point has proved itself fruitful; and Montague's (1974) intensional model of language has been followed by a number of elaborations and modifications. There were also alternative intensional models (due to Cresswell, 1973, Tichý, 1978, and others) proposed partly or wholly independently of Montague's approach. Then there followed modified, "hyperintensional" models of semantics attempting to improve on the intensional model especially to make it capable of adequately analyzing the so-called "propositional attitude reports" (see Bigelow, 1978, or Lewis, 1972). To these we can count systems based on the so called structured or Lewis-type meanings (Cresswell, 1985), Tichý's (1988) theory of constructions, Barwise's & Perry's (1983) situation semantics, etc. There then followed models reflecting the "dynamic" aspect of natural language, such as Kamp's (1981) DRT or various models based on dynamic logics (van Benthem, 1997). The early state of the art was excellently surveyed by van Benthem & ter Meulen (1996).

# 3 Extensional model of meaning: Frege's maneuver exploited to the bone

## 3.1 Principles of etensional semantics

If what we are after is capturing meanings as objects, then the category of expressions that becomes immediately salient for us is that of names, for these appear to stand for objects. Especially the function of proper names seems to be exhausted by representing objects quite straightforwardly[20]; but even names in a broader sense (perhaps singular nominal phrases) can be clearly seen as representations of objects. If we decide to identify the meanings of all such names, in the broad sense of the word, with the objects named by them, we are on the track of the extensional model of meaning. So let us assume, together with Frege that

(I) the meaning of a name (not just proper name, but plus/minus any singular noun phrase) is the object it names (or, as we will sometimes call it, *an individual*). For example, the meaning of *Eco* and *author of the novel The Name of the Rose* is the person Eco, the meaning of *morning star, evening star, Venus, brightest star in the morning sky* and *brightest star in the evening sky* is the planet Venus, etc.

Frege, however, argues that it is not only names, but also sentences that stand for objects, though sentences stand for objects of a special kind. In particular, he argues that

(II) the meaning of a sentence is its truth value. The meaning of sentence (9) is therefore the truth value **Tr**, while the meaning of sentence (10) is the truth value **Fa**.

(9) *Eco is a writer*

---

[20] Though many semanticists propose that proper names have some meaning over and above this.

(10) *Eco is an actor*

While assumption (I) captures the natural idea of names as naming things, assumption (II) is far from natural. Why should we declare the meaning of a sentence to be its truth value and not, say, something like a "situation" expressed by this sentence? However, (II), as Frege pointed out, is in a sense a consequence of (I).

The point is that if we accept the principle of compositionality, we thereby also accept the principle of intersubstitutivity of synonyms (see §1.5). Given this, we can argue in the following way (this particular version of the argument comes from Church, 1956, p. 25):

Consider sentences (11a) – (11d).

(11a) *Walter Scott is the author of Waverley*

(11b) *Walter Scott is the man who wrote twenty-nine Waverley novels altogether*

(11c) *The number, such that Walter Scott is the man who wrote that many Waverley novels altogether, is twenty-nine*

(11d) *The number of counties in Utah is twenty-nine.*

The sentence (11b) arises from (11a) by replacing the term *the author of Waverley* with the term *the man who wrote twenty-nine Waverley Novels altogether*. Since both terms obviously refer to the same person, they have the same extension, and so if we accept (I), we must conclude that (11b) has the same meaning as (11a). (11c) is obviously just a reformulation of (11b), so it is synonymous with (11b), and is therefore synonymous with (11a). And because (11d) results from (11c) again by replacing a phrase with a coreferential one, (11d) has the same meaning (extension) as (11c), and hence (11d) is synonymous with (11a). However, the only thing (11a) and (11d) really have in common is the truth value (both are true). And because more general considerations, Frege maintained, indicate that the truth value is never affected by replacing names with other names with the same extension, (II) is forthcoming. This reasoning then seems to lead to the conclusion that if we accept (I) and the principle of compositionality, we must also accept (II).

We have already seen that Frege, who initiated the extensional approach to meaning, did not think that such an extensionally understood meaning would actually explicate meaning in an intuitive sense. (It would certainly be more difficult to accept the conclusion that all true sentences are, as in the extensional model, synonymous.) It is for this simplicity that the extensional model is suitable to demonstrate some elementary semantic mechanisms, which are then used in more complex and more adequate semantic models.

## 3.2 Subject and predicate

Accepting the principles outlined in the previous section, let us start to build our first, extensional model of language. A typical simple sentence consists of a subject (nominal phrase) and a predicate (intransitive verbal phrase). We will reflect this in our model, but to distinguish the expression of the language of the model from natural language, we will use a slightly modified terminology: the expressions corresponding to sentences will be called *statements*; and those corresponding to subjects will be called *terms*. Only the expressions corresponding to predicates will be called *predicates* – out of a lack of an alternative. Hence our model, the extensional model of language, starts from three basic categories of expressions: *statements, terms* and (*unary*) *predicates*. A statement typically consists of a term and a predicate; within our model, we will capture this connection by specifying the term in parentheses after the predicate. We can therefore rewrite (9) as (9′) and (10) as (10′).

(9′) <u>writer(Eco)</u>

(10′) <u>actor(Eco)</u>

By way of generalization, we can formulate the first syntactic rule of our formal language model:

[A] If $P$ is a predicate and $T$ is a term, then $P(T)$ is a statement

(9′) and (10′) are not really sentences, they are statements (sometimes also called formulas) that we understand as certain "normalized" versions of sentences; these statements are elements of our model. Compared to sentences, they show more explicitly their syntactic structure, which is relevant from the point of view of semantics. We will (as we have already started) underline the "normalized" counterparts of natural language words.

We said that within the framework of extensional semantics, the meaning of a term is considered to be an object that is named by this term. The set of all those objects that can be named by the terms of our language, that is, the set of all those objects that we are able to talk about in terms of our language, we call the *universe of discourse*.

The model we are building will also, of course, contain meanings, or their explananda, which are in this case extensions. I will use the technical term *denotation* instead of the term *meaning* when talking about the model. If we represent the denotation of an expression by enclosing the expression in the symbols '$\|$' (as we did already in the previous chapter), we can write

$\|\underline{Eco}\| = $ Eco

$\|\underline{Schwarzenegger}\| = $ Schwarzenegger

$\|\underline{morning\ star}\| = $ Venus

...

The universe of discourse includes all objects that can be denotations of our terms; it is therefore a set containing Eco, Schwarzenegger, Venus and all the other individuals we can talk about.

We identify the denotation of a statement, as we have said, with its truth value; we therefore consider it as an element of the set $\{\boldsymbol{Tr}, \boldsymbol{Fa}\}$. So

$\|\underline{writer(Eco)}\| = \boldsymbol{Tr}$

$\|\underline{actor(Eco)}\| = \boldsymbol{Fa}$

$\|\underline{actor(Schwarzenegger)}\| = \boldsymbol{Tr}$

...

What should we see as the denotation of a predicate? We stated that the meaning of an expression should be determined by the meanings of parts of that expression (principle of compositionality); thus, in our particular case, the denotation of a statement created according to rule [A] should be determined by the denotation of its term and that of its predicate. The statement (9′) is the result of a combination of the term *Eco* and the predicate *writer*, so the denotation ‖*writer(Eco)*‖ (which is the truth value of *Tr*) must come from some combination of denotations ‖*Eco*‖ (which is the person Eco) and ‖*writer*‖.

Here we accomplish what we dubbed, in the introductory chapter, Frege's maneuver. We can look at a predicate as something that together with a term creates a statement, which therefore "makes statements" from terms (so far we are at the level of syntax, i.e. expressions, not their meanings). The predicate *writer*, from this point of view, "makes" from the term *Eco* the statement *writer(Eco)*, from the term *Schwarzenegger* the statement *writer(Schwarzenegger)*, etc.; we can therefore look at it in this sense as a function that we can symbolically write in the following way[21]:

*Eco* ⟶ *writer(Eco)*

*Schwarzenegger* ⟶ *writer(Schwarzenegger)*

...

If we do analogous reasoning at the level of semantics (i.e. if we replace the expressions with their denotations), we conclude that the denotation of the predicate *writer* transforms the denotation of the term *Eco* into the denotation of the statement *writer(Eco)*, the denotation of the term *Schwarzenegger* into the denotation of the statement *writer(Schwarzenegger)*, etc. We can therefore look at it as a function that we can write as:

‖*Eco*‖ ⟶ ‖*writer(Eco)*‖

‖*Schwarzenegger*‖ ⟶ ‖*writer(Schwarzenegger)*‖

---

[21] We will sometimes specify functions in this graphic way. The notation $A \longrightarrow B$ means, as is perhaps obvious, that the function in question assigns the object (value) *B* to the object (argument) *A*.

...

that is, as a function that assigns the truth value ***Tr*** to Eco, the truth value ***Fa*** to Schwarzenegger, etc. (generally truth values to individuals):

Eco $\longrightarrow$ ***Tr***

Schwarzenegger $\longrightarrow$ ***Fa***

...

Here we should stress that Frege, who was responsible for the maneuver, did not see functions as objects. According to him, functions, by their nature, were "unsaturated", and to form an object they must come to be "saturated". This happens when we apply function to an object as its argument – the function becomes saturated and becomes an object – the value of the function. The set of <argument, value> pairs is *not* the function itself, it is called the course-of-values of the function and it is, unlike the function, an object. However, since, after Frege, the distinction between a function and its course-of-value slowly died away in mathematics, we can directly see functions – and especially concepts – as (set-theoretical) objects. Thus, we have denotations for all the three basic categories of expressions of our incipient model.

Alternatively, we could take the denotation of the predicate as a set of individuals, i.e. *a subset of the universe of discourse* – specifically, the set of those objects of which the predicate is true (so we can find it in many textbooks of predicate logic). Then it holds that a statement that consists of a term and a predicate is true (its denotation is ***Tr***) just when the object denoted by its term is an element of the set denoted by its predicate. Thus, for the denotation of the predicate <u>writer</u> (which, of course, can be understood as capturing the phrase (*to be a*) *writer*[22]), we take the set of all writers; and statement (9′) is true just when the denotation of the term <u>Eco</u>, i.e. the person Eco, is an element of the denotation of the predicate <u>writer</u>, i.e. the set of writers. The sentence (9) is thus reconstructed as true

---

[22] The verb *be* can be considered a legitimate part of the predicate, or just an aid to the realization of the term with the predicate.

just when Eco is a writer; and that's obviously how it's supposed to be. In that case, we can write

$\|\underline{writer}\|$ = {Eco, Rushdie, ...}

$\|\underline{actor}\|$ = {Schwarzenegger, diCaprio, ...}

What is the difference between this answer to the question of what is the denotation of a predicate and the answer given above, namely that its meaning is a function that assigns truth values to individuals? Is the "real" denotation the function from the universe to truth values, or a subset of the universe? It is easy to see that there is no significant difference between these answers, and that we can therefore only see them as different variants of a single answer. There is a simple one-one relationship between the subsets of the universe and the functions from the universe to the truth values, and we can thus largely interchange one with the other for our purposes. The function from the universe to the truth values can always be seen as the definition of a certain set: it divides the universe into two groups, the elements to which it assigns *Tr*, and the ones to which it assigns *Fa*[23]; and we can see it as separating those elements that belong to the defined set from those that do not belong to it. The function assigning *Tr* to all writers and *Fa* to all other elements of the universe will therefore be considered essentially the same as the set of all writers. Given this, we will continue to view the denotation of the predicate as a subset of the universe, as well as a function from the universe to truth values, as needed.

In general, we can represent each subset $S^*$ of a given set $S$ by its *characteristic function* $F_{S^*}$ from $S$ to $\{\textbf{\textit{Tr}}, \textbf{\textit{Fa}}\}$ such that $F_{S^*}(x) = \textbf{\textit{Tr}}$ just when $x \in S^*$. Conversely, every function $F$ from $S$ to $\{\textbf{\textit{Tr}}, \textbf{\textit{Fa}}\}$ can be understood as a characteristic function of some subset of $S$, namely the set $S_F = \{x \in S \mid F(x) = \textbf{\textit{Tr}}\}$.

However, if we understand the denotation of a predicate as a function from the universe to truth values, the denotation of a statement created

---

[23] This presupposes that the functions we talk about are *total*, i.e. that they assign the value *Tr* or *Fa* to *every* element of the universe.

according to rule [A] will simply be an application of the denotation of its predicate to the denotation of its term. The denotation of the statement (9') is therefore the result of the application of the above-mentioned function to Eco, i.e. the value ***Tr.*** In general, if $P$ is a predicate and $T$ is a term, then

[A'] $\|P(T)\| = \|P\|(\|T\|)$.

## 3.3 Functions

Let me emphasize that a significant part of the reason why set theory seems so well suited to this approach to meaning is that functions are understood as sets. For within modern mathematics, a function is understood as a set of ordered pairs where an ordered pair can, as we have already seen, be grasped as a special kind of set.

We have also already mentioned that Frege, for example, distinguished between a function as such and a course-of-values of the function. While the function itself was not an object for him, since it was something fundamentally "unsaturated" (i.e. incomplete), its course-of-values was an object – it was de facto the set of ordered pairs that the function is for mathematicians today. For Frege, then, the notations

$x + x + 1$

and

$1 + 2.x$

indicated two different functions, sharing the same course-of-values.

The important thing, of course, was that while traditionally functions were understood as a matter of numbers (i.e. typically they assigned numbers to numbers), for Frege and other modern mathematicians they are understood more generally – we can have a function that assigns to each day in a person's life what that person had for breakfast that day, i.e. a roll on one day, a cake on another, and nothing on another when they overslept. Hence, when we are operating within set theory, a function can assign any kinds of sets to sets (it can assign, for example, also functions to functions).

But here we come to a point which is worth discussing in greater detail than is usually the case. In fact, though functions are understood as sets, the objects they are built of (that is the objects which the function assigns to other objects) are not necessarily sets. We saw that, following Frege and Carnap, we took the denotation of a predicate to be a function that assigns truth values to objects of the universe, where neither the truth values nor the elements of the universe were explicitly required to be sets.

The reason, we can say, is that we operate within the version of set theory which is usually called the set theory with ur-elements and which is not the kind of set theory mathematicians operate with. This kind of set theory assumes that there are some elements that are pre-given and over which the sets are formed. (Mathematicians, in contrast to this, usually work within so-called pure set theory, in which no such ur-elements obtain.)

Including ur-elements in set theory seems quite natural. It would seem that first we must have some objects from which we can form sets (and then we can perhaps form sets also from the already formed sets), the objects being necessary and being an integral part of the theory. (Within pure set theory the role of such elementary objects, like e.g. numbers, are played by sets formed by means of the empty set, such as $\emptyset$, $\{\emptyset\}$, $\{\emptyset,\emptyset\}$, $\{\emptyset,\{\emptyset\}\}$ etc.)

The trouble with ur-elements is that when we treat them too carelessly, we may come to build a theory which looks like a purely mathematical theory into which the ur-elements smuggle an empirical dimension. Consider, for example, the denotation of the predicate *dog*, which would be the function mapping all objects of our universe on truth values, according to whether the object in question is a dog or not. This may seem to be a distinctively mathematical object – a set of ordered pairs. However, it contains empirical objects, which are of an inherently uncertain nature. So we have a set of ordered pairs but with no determinate criterion of which pairs belong to it. Is, for example, this kind of creature really a dog and does it belong to the function together with the value ***Tr***? The empirical world need not deliver a unique answer to such a question.

## 3.4 Quantification

Thus, our rule [A] combines terms with predicates into statements. However, it seems that the denotation we have given to terms does not allow us to reasonably capture all kinds of expressions that occur as subjects in natural language sentences, and therefore rule [A] cannot be understood as a general expression of a combination of all kinds of subjects with all kinds of predicates. For example, consider the following sentences

(12) *Someone is a writer*

(13) *Everyone is a writer*.

The subject of sentence (12), the term *someone*, or the subject of (13), *everyone*, would certainly not be reasonably considered as naming one particular individual. This means that not all the subjects of natural language, indeed, far from it, would be reasonably seen as terms in our sense – that is, as the names of elements of the universe.

Thus, it seems that terms like *someone* or *everyone* must be placed in a different category than terms; we will call them *quantifiers*. Then, however, a predicate can be combined into a statement not only with a term, but also with a quantifier. Therefore, in addition to the syntactic rule [A], we need another, analogous rule, which will express the composition of a predicate into a statement not with a term, but with a quantifier. But before we formulate such a rule, let's think about what could be taken as the denotation of a quantifier in our model.

Here, as in the case of predicates and unlike in the case of terms, we cannot rely on any clear intuition. Recall how we proceeded with predicates: we assumed that the denotation of a predicate should be something that (according to the principle of compositionality) gives us, together with the denotation of a term (individual), the denotation of the resulting statement (truth value), so we simply identified this denotation with a function that assigns truth values to objects – the denotation of a statement formed by a predicate and a term is now obtained simply as the denotation of its predicate applied to the denotation of its term. Could we do the same for quantifiers? A quantifier is, as we have just said, something that forms, just as a term, a statement along with a predicate; and the denotation of such a statement must then be somehow determined by the denotation of its predicate and the denotation of its quantifier. Apparently, however, we can no longer do this so that the denotation of this statement (truth value) is the result of applying the denotation of its predicate (function from individuals to truth values) to the denotation of its quantifier – because then the denotation of the quantifier would have to be an individual, and that is what we didn't want to put up with. But what if we did it so that, conversely, the denotation of such a statement were the result of applying the denotation of its quantifier to the denotation of its predicate? In such a case, the denotation of the quantifier would have to be a function that assigns the denotations of predicates (functions from individuals to truth values) the denotations of the respective statements (truth values). So why not go for this?

So let the denotation of the quantifier be the function that assigns truth values to the functions that assign truth values to individuals. Since "classification" functions that assign truth values to some kind of objects can be viewed, as we said in the previous section, as sets of the objects to which the functions assign *Tr*, we can talk about sets of individuals instead of functions that assign truth values to individuals. Thus, we can say that the denotations of quantifiers will be functions that assign truth values to sets of individuals. And if we use the same kind of reformulation once again, we can say that they will be sets of sets of individuals.

The statement which is created by combining a predicate and a quantifier will be understood as the application of its quantifier to its predicate, and the relevant rule can be formulated as follows:

[B] If $P$ is a predicate and $Q$ is a quantifier, then $Q(P)$ is a statement.

The denotation of such a statement will then be

[B'] $\|Q(P)\| = \|Q\|(\|P\|)$.

Therefore, if the quantifier that, within our model, corresponds to the expression *someone* is Σ, sentence (12) will correspond to the statement

(12′) Σ(*writer*)

and it will be the case that

$\|\Sigma(writer)\| = \|\Sigma\|(\|writer\|)$.

What will be the denotation of the quantifier Σ? We have said that it will be a function that assigns truth values to functions that assign truth values to individuals, that is, a set of sets of individuals; but which such function will it be? The sentence (12) is true if there is a writer, that is, if the denotation of the predicate *writer* is a nonempty set. If no writer existed, that is, if the set of writers (which is the meaning of the predicate *writer*) were empty, this statement would be false. That is, the denotation of Σ will be the set of all nonempty subsets of the universe, or a function that assigns *Tr* to every function that assigns *Tr* to at least one individual.[24] Similarly, if we introduce the quantifier Π as the formal opposite of the term *everyone*, its denotation will be that set of subsets of the universe which contains a single subset, namely the one that is identical with the whole universe; which is to say, it will be a function that assigns *Tr* to only the function that assigns *Tr* to every individual.

Thus, we can say that we take a quantifier as a kind of "predicate of predicates" that attributes some property to the denotation of the predicate to which it is applied (analogously to a predicate, which attributes some property, such as writing, to the denotation of the term to which it is applied, such as the individual Eco). For example, Σ attributes non-emptiness to the denotation of the associated predicate: Σ(P) "says" that the set $\|P\|$ contains at least one element. Similarly, Π attributes the denotation of the associated predicate "universality": Π(P) "says" that the set $\|P\|$ contains all the elements of the universe.

---

[24] Here, however, we completely ignore the fact that *someone* refers in natural language only to humans; therefore, our quantifier Σ corresponds more to something like *someone or something*.

## 3.5 Negation and propositional connectives

In the previous sections, we have accounted for the most basic syntactic-semantic nexus of English (and presumably of any natural language), the combination of subject and predicate into a simple sentence. Now we move to something perhaps less central from the viewpoint of natural language but central from the viewpoint of logic on which the enterprise of formal semantics largely rests, which is the operators of propositional logic and their counterparts in English. It turns out that even the explication of their meanings (which is well known from propositional logic) can be seen as based on Frege's maneuver.

Sentence (14) can be considered to consist of two parts: the *is not true* part and sentence (9). In sentence (15), which is apparently essentially just a reformulation of (14),[25] the role of the first of these parts is taken over by the word *not*.

(14) *It is not true that Eco is a writer*

(15) *Eco is not a writer*

If we denote negation, whether it is actually expressed as in (14) or as in (15), by the symbol $\neg$, then we have

If $S$ is a statement, then $\neg S$ is a statement.

By way of generalization we can formulate a new rule, calling the expressions of the kind of negation propositional operators:

[C] If $S$ is a statement and $O$ is a propositional operator, then $OS$ is a statement.

In this way we can understand negation as a certain "predicate of statements". Just as normal predicates combine with terms into

---

[25] It is necessary to realize that negation in natural language is a phenomenon much more complex than it appears when it is schematized, as it is here by the means of elementary logic. It does not only apply to whole sentences, we can also negate only parts of sentences or individual words, etc. Hence using $\neg$ as its capture is a grave oversimplification.

statements, ¬ combines with statements into statements. If we concluded above, using Frege's maneuver, that the denotations of predicates are functions assigning truth values (denotations of statements) to objects (denotations of terms) and that the denotations of quantifiers are functions assigning truth values (denotations of statements) to sets of objects (denotations of predicates), then we can derive, by parity of reasoning, that it may be reasonable to take the denotation of negation as a function that assigns truth values (denotations of statements) to truth values (denotations of statements). Since the statement ¬$S$ appears to be true just when the statement $S$ is false, the meaning of ¬ is the function that assigns the value *Fa* to the value *Tr* and the value *Tr* to the value *Fa*; therefore

(T¬) ‖¬‖ =

   *Tr* ⟶ *Fa*

   *Fa* ⟶ *Tr*

The meaning of the negated sentence will then be

‖¬$S$‖ = ‖¬‖(‖$S$‖),

i.e. more generally

[C'] ‖$OS$‖ = ‖$O$‖(‖$S$‖).

If we "calculate" the meaning of sentence (15), we see that we get the correct value.

‖*Eco is not a writer*‖ =

‖¬*writer(Eco)*‖ =

‖¬‖(‖*writer(Eco)*‖) =

‖¬‖(*Tr*) = *Fa*

Regarding the joining of pairs of sentences, the basic cases will be *conjunction, disjunction* and *implication*, more or less corresponding to the English connectives *and, or* and *if-then*. For these three types of connectives we will use, as is common in logic, the symbols ∧, ∨ and →. These symbols are called *propositional connectives*. So it is the case that

If $S_1$ and $S_2$ are statements, then $(S_1 \wedge S_2)$, $(S_1 \vee S_2)$ and $(S_1 \rightarrow S_2)$ are also statements

and we have another syntactic rule

[D] If $S_1$ and $S_2$ are statements and C is a propositional connective, then $(S_1\ C\ S_2)$ is a statement.

Just as we have concluded that the denotation of ¬ as a function that assigns truth values to truth values, we conclude that it is reasonable to take these connectives as denoting functions assigning truth values to *pairs* of truth values; so it will be

[D'] $\|S_1\ C\ S_2\| = \|C\|(\|S_1\|, \|S_2\|)$.

The conjunction of two statements is true when both of the statements are true, so the denotation of $\wedge$ is the following function:

(T$\wedge$) $\|\wedge\| =$

> ***Tr, Tr*** $\longrightarrow$ ***T***
> 
> ***Tr, Fa*** $\longrightarrow$ ***Fa***
> 
> ***Fa, Tr*** $\longrightarrow$ ***Fa***
> 
> ***Fa, Fa*** $\longrightarrow$ ***Fa***

So

$\|$ Eco is a writer and Schwarzenegger is an actor $\| =$

$\|\underline{writer(Eco) \wedge actor(Schwarzenegger)}\| =$

$\|\wedge\|(\|\underline{writer(Eco)}\|, \|\underline{actor(Schwarzenegger)}\|) =$

$\|\wedge\|(***Tr, Tr***) = ***Tr***$

The disjunction of two statements is apparently true just when at least one of the two statements is true. Therefore:

(T∨) $\|\vee\|=$

*Tr, Tr* ⟶ *Tr*

*Tr, Fa* ⟶ *Tr*

*Fa, Tr* ⟶ *Tr*

*Fa, Fa* ⟶ *Fa*

The implication is slightly less transparent. To shed some light on the situation, let's look at how the connective *if-then*, which is its counterpart in natural language, behaves. A sentence formed from two sentences connected by *if-then* is apparently true if both the sentences are true and is false if the first of the sentences is true and the second is false. This is because if Eco is a writer, then the phrase *If Eco is a writer, then Schwarzenegger is an actor* is true just when Schwarzenegger really is an actor. But what if the first sentence is false? What would be the truth value of the sentence *If Eco is a writer, then Schwarzenegger is an actor* if Eco were not a writer? In a sense, there is no reason to consider this sentence false: this can be seen in the example of statements such as *If Eco is an actor, then I am the pope*: if I am sure that the first of the sentences (*Eco is an actor*) is false, then I can take anything else (perhaps a sentence as obviously false as *I am the pope*) and the result will be true; that is, if the first sentence of the implication is false (has a truth value of *Fa*), the whole implication is true (has a value of *Tr*), regardless of the truth value of the second sentence.[26] So we have

---

[26] It should be noted, however, that a connection *if-then* is often used in natural language in other ways as well, namely as an expression of something as a causal connection. (A sentence like *If it rains, then it's wet* is not normally understood to mean that not raining would suffice to make it true.) Therefore, the operator →, again, is an oversimplification, it cannot be said to be a straightforward capture of this connection. This is the reason why modern logic abounds with attempts to replace this kind of implication with a "better" one.

(T→) $\|\to\|$ =

*Tr, Tr* ⟶ *Tr*

*Tr, Fa* ⟶ *Fa*

*Fa, Tr* ⟶ *Tr*

*Fa, Fa* ⟶ *Tr*

## 3.6 Basic extensional model

In the previous sections, by reconstructing some of the syntactic rules of natural language, we arrived at a certain simple formal language. Let's summarize now what this language looks like. We will consider it as our initial extensional model of natural language; we will call it $L_E$. Let us first recapitulate the syntactic structure (for rules, we state in square brackets after the rule number under which we introduced the rule above)[27]:

There are six categories of expressions, namely *terms* (**T**), *predicates* (**P**), *propositional operators* (**O**), *propositional connectives* (**C**), *quantifiers* (**Q**) and *statements* (**S**). The vocabulary of the model consists of the following words: an unlimited number of simple expressions of the categories **T** and **P**; ¬ of the category **O**; ∧, ∨ and → of the category **C**; and Σ and Π of the category **Q**. In addition, we assume that we have auxiliary, *syncategorematic* symbols, namely parentheses.

As for the syntactic rules, there are four of them:

[A] If *P* is a predicate and *T* is a term, then *P(T)* is a statement.

[B] If *Q* is a quantifier and *P* is a predicate, then *Q(P)* is a statement.

[C] If *O* is a propositional operator and *S* is a statement, then *OS* is a statement.

[D] If *C* is a propositional connective and $S_1$, $S_2$ are statements, then $(S_1 \, C \, S_2)$ is a statement.

---

[27] For a more rigorous description of the model, see §8.1.

Before we recapitulate the assignment of meanings to the terms just defined, let us introduce some notation conventions. If M and N are two sets, then the symbol [M ⇨ N] will denote the set of all functions from M to N (i.e. certain sets of ordered pairs of elements of M resp. N). The symbol B will denote the set of two truth values, i.e. the set {*Tr*, *Fa*}. [M×N ⇨ O] will then denote the set of all two-argument functions from the sets M and N to the set O (thus, mathematically speaking, the function from the Cartesian product M×N to O); so, for example, the meanings of propositional connectives will be elements of [B×B ⇨ B]. i.e. functions assigning truth values to pairs of truth values. As we discussed above, we will often identify the "classificatory" function $f$ from some set M to the set B (that is, the element of the set [M ⇨ B]) with a subset of M, namely the set of all those elements $m \in M$ for which $f(m) = $ ***Tr***.

We can now put together the semantic part of the definition of the language $L_E$. The denotations of simple expressions are as follows: The denotation of a term is an element of the given universe U. The denotation of a predicate is a function from U to B, i.e. an element of [U ⇨ B] (or a subset of U, i.e. a set of individuals). The denotation of a propositional operator is a function from B to B, i.e. an element of [B ⇨ B]; the denotation of the operator ¬ is the function given by the above table (T¬). The denotation of a propositional connective is a function from B×B to B, i.e. the element of [B×B ⇨ B]; the meaning of the operators ∧, ∨ and → are those functions which are given by the above tables (T∧), (T∨) and (T→).

The denotation of a quantifier is a function from [U ⇨ B] (sets of individual) to B, i.e. an element of [[U ⇨ B] ⇨ B] (sets of sets of individuals); the denotation of the quantifier Π is the function that assigns the value ***Tr*** to a function $f$ just when $f(i) = $ ***Tr*** for every individual $i \in U$ (or it is such a set of sets of individuals which contains a single set of individuals, namely the one that is identical with U, i.e. contains all the individuals of the universe); the denotation of Σ is the function that assigns the value ***Tr*** to the function $f$ just when $f(i) = $ ***Tr*** for at least one individual $i \in U$ (the set of sets of individuals that contains all those sets of individuals that are nonempty, i.e. that contain at least one individual).

The denotation of the statement is a truth value, i.e. an element of B.

The rules for computing the denotations of compound expressions from those of their components are now as follows:

[A'] $\|P(T)\| = \|P\|(\|T\|)$.
[B'] $\|Q(P)\| = \|Q\|(\|P\|)$.
[C'] $\|OS\| = \|O\|(\|S\|)$.
[D'] $\|S_1 \, C \, S_2\| = \|C\|(\|S_1\|, \|S_2\|)$.

## 3.7 Fregean quantification

The language $L_E$ is a very primitive model, into which it is still not possible to absorb too much natural language. Consider, for example, the sentence

(16) *Eco is a writer and an actor.*

This sentence apparently consists of the subject *Eco* and the predicate (*being*) *a writer and an actor* (which fits into our model as a primitive predicate), but its predicate *writer and actor* apparently consists of conjunctively connected parts *writer* and *actor*. To be able to capture this in our model as well, we would have to extend it with the possibility of conjunctively (or disjunctively, etc.) connecting not only statements but also predicates.

However, if it were only sentences like (16), the question would be whether we really need something like that: sentence (16) says the same thing as

(17) *Eco is a writer and Eco is an actor,*

which we can already capture in our model (as a conjunction of two statements, each of which is a combination of a term and a predicate). Thus, one might think that we would consider sentence (16) as a dispensable paraphrase. However, it is different with the sentence

(18) *Someone is a writer and an actor.*

Sentence (18) obviously says something else than the sentence

(19) *Someone is a writer and someone is an actor*

and therefore we cannot consider it as a dispensable paraphrase.

A straightforward way to include sentences of this kind in our model would probably be to introduce a rule that says that if $P_1$ and $P_2$ are predicates then $(P_1 \wedge P_2)$ is also a predicate. (Note, however, that such a rule could no longer be semantically such that the denotation of the predicate created by it is the result of applying the denotation of one of its parts to the denotations of other parts. None of the expressions that enter this rule has a denotation that can be applied to those of the others.) But there is another way. It is not natural from our point of view, but it will bring our model closer to standard logic, which is why we will now set out on it.

This way is based on the fact that instead of combining predicates, there is a possibility of "disguising predicates as statements", compounding the resulting statements, and then removing the disguise to obtain a compound predicate. The talk of "disguising", of course, is metaphoric, and those who introduced the method would not endorse it; but I think that from the current viewpoint it is illuminating. The fact is that this method originated as a by-product of Frege's and Russell's approach to quantification.

Bertrand Russell, whose logical analysis of sentences of type (10) became at the beginning of the twentieth century a paradigm of modern logic's approach to language and remained so for a long time, relied on Frege's theory of quantification in the analysis of the sentences (Russell, 1905). The result was that, for example, the sentence (10) was analyzed as *the sentence 'x is a writer and actor' is sometimes* (that is: *for some x*) *true*, which, when expressed in the notation of modern logic, gives us $\exists x(\underline{writer}(x) \wedge \underline{actor}(x))$. To follow this path, we must first explore a way to get the Fregean quantification into our model.

Frege's notion of quantifier is based on the idea that we can omit some part of a sentence (replace it with an empty symbol, such as $x$) and then consider what truth values this sentence will give if we fill the vacancy in different ways. For example, if we omit the exression *Eco* in the sentence (16), we get (20); if we omit the expression *Schwarzenegger* in (21), we get (22).

(20) *x is a writer and actor*

(21) *If Schwarzenegger is human, then Schwarzenegger is mortal*

(22) *If x is human, then x is mortal.*

(20) and (22), however, are not sentences, they are mere *sentential schemes* (sometimes we also talk about *matrices*), which can become sentences only when we fill the vacancies in them. The propositional scheme (20) gives a true sentence if we replace *x* with *Eco* but gives a false sentence if we replace *x* with *Schwarzenegger*. The propositional scheme (22), on the other hand, becomes a true sentence, no matter what we substitute for *x*.

We can formulate various statements about sentential schemes and what happens when we substitute different expressions into them. For example, we can say that something can be substituted into a sentential scheme to give a true sentence (this applies to schemes (20) and (22)); or we can say that whatever we put into such a scheme, we will always get a true sentence (this is the case (22)). We will use the symbol $\exists$ to express the first of these types of statements, as is usual in logic, and $\forall$ to express the second. So if we write $\exists x(x$ *is a writer and an actor*), we can read it as *there is a term such that if we substitute it for x into* (20), *we get a true sentence*, or simply there is an *x such that x is a writer and an actor*; and we write $\forall x(x$ *is a writer and an actor*) for *for every term, if we substitute it for x into* (20), *we get a true sentence,* or simply *for every x, x is a writer and an actor*. We will talk about $\forall$ *as a universal* quantifier, and $\exists$ then as an *existential* quantifier.

More specifically, there are two different ways of understanding Fregean quantifiers. The one we have just described is called *substitutional*. In addition, there is an even more common way called *objectual* or *denotational,* where instead of imagining that we replace *x* with different terms, we imagine that we let *x* denote different things. $\exists x(x$ *is a writer and an actor*) is then to be read as *there is such an element of the universe that if we consider it to be denoted by x, it ill make* (20) *true*. If each element of the universe has a name (that is, if there is a term that denotes it), the two understandings are equivalent; however, if there is an element in the universe that does not have a name, they may not be equivalent. An existentially quantified sentence may be true in the objectual understanding of quantification (because there is an object that makes it true) and false in the substitutional (because there is no term that would make it true).

In this way, we seem to have moved away from our subject matter because we stopped talking about the meanings of sentences and their parts, and instead we began to deal with statements about how sentential schemes are made into sentences. The point of our detour is that what we have formulated as statements about schemes can (according to Russell) be understood as analyses of quantified sentences of natural language. So to say that *there exists x such that x is a writer and an actor* is nothing other than to say that *someone is a writer and an actor,* and to say that *for every x it holds that if x is human, then x is mortal,* is the same as saying that *every human is mortal* or simply that *humans are mortal.*

The point is that sentential schemes can be considered as "pseudopredicates". Just as a predicate yields a statement if it is combined with a term, a sentential scheme yields a sentence if its variable is replaced by a term. Hence, here is the recipe as to how to conjoin predicates: We take two predicates, say <u>writer</u> and <u>actor</u>, use a variable to produce sentential schemes, <u>writer</u>($x$) and <u>actor</u>($x$), conjoin them and we have a complex sentential scheme <u>writer</u>($x$)∧<u>actor</u>($x$), aka a pseudopredicate. If now quantifiers can be applied to it, we have solved the problem of the analysis of sentence (23). And this is what Fregean quantifiers can indeed do.

In Section 3.4, we introduced a (transparent) syntactic rule [B] for associating quantifiers with predicates and provided it with the appropriate semantics; however, the adoption of Fregean quantifiers requires that this rule be replaced by something else, much less transparent. It requires the introduction of the possibility to create sentential schemes ("statements" with variables in place of terms) and the introduction of a rule for creating statements from quantifiers plus sentential schemes. If we introduce into our language new kinds of expressions, variable terms, or simply *variables,* we can formulate rules [B1] and [B2]:

[B1] If $P$ is a predicate and $x$ is a variable, then $P(x)$ is a statement,

[B2] If $S$ is a statement and $x$ is a variable, $\exists x(S)$ and $\forall x(S)$ are statements.

However, if we actually accepted these rules it would mean that our formal language, which we are building as a model of natural language, would move away from natural language in an undesirable way. This would mean that we would have to include expressions that have no obvious equivalents in natural language (variables, statements that are in fact schemes); our model would thus become in an unpleasant way unlike what it is supposed to model. Therefore, we will choose a different path: we combine [B1] and [B2] into a single rule [B*] and thus *de facto* reach the point that the variables and schemes grow into the role of aids needed only to express this rule and are not full-fledged expressions of our formal language. If we use the symbol $S^{T \leftarrow x}$ as a designation of the scheme that arises from $S$ when the term $T$ *is* replaced by the variable $x$, we can formulate the rule [B*] as follows:

[B*] If $S$ is a statement and $T$ is a term, then $\exists x(S^{T \leftarrow x})$ and $\forall x(S^{T \leftarrow x})$ are statements.

To appreciate the difference between [B1] and [B2], on the one hand, and [B*] on the other, we must be properly aware of the difference between the formal means we introduce *into the language* we are building and the formal means we use *to deal with this language.* If we added rules [B1] and [B2] to a language made up of terms, predicates and statements and syntactic rule [A], we would have to add variables (because they are an input for rule [B1]) and schemes (because they are produced by [B1] and are a kind of statement). Thus, expressions like $x$ and writer($x$) would get into the language we are building. If, on the other hand, we add the rule [B*], we are spared this: the input of this rule is statements like *writer*(*Eco*), and its output is again statements like ∃*x*(*writer*(*x*)). (Let's not be fooled by the fact that the output statements contain variables – they are not parts of the vocabulary of the language, here they are just technical tools of indicating which quantifiers are connected to which places in formulas.)

As a result, we deviate from what is common in logic. There we usually introduce quantification with rules of type [B1] and [B2] and look at variables as syntactically fully-fledged terms and schemes as syntactically fully-fledged statements; and we get a formal language that is extremely simple and flexible at the same time. However, here we are primarily concerned with maintaining the correspondence between the formal language being created and the natural language because we are building the former as a model of the latter. Thus, unlike logic, for us, variables are nothing more than irrelevant auxiliary symbols such as parentheses.

So we will now capture (24) as (24'), and (25) as (25'):

(24) *Someone* (or *something*) *is a writer*

(24') ∃*x*(*writer*(*x*))

(25) *Everyone* (or *everything*) *is a writer*

(25') ∀*x*(*writer*(*x*)).

Note the fundamental difference between rules [B] and [B*]. While rule [B] directly reflects one of the basic syntactic operations of natural language (that is, combining a certain kind of subject with predicates), rule [B*] is more widely applicable: with its help, for example, we can create a statement ∀*x*(*human*(*x*)→*mortal*(*x*)), which does not correspond to any statement we could make with rule [B]. The reason is that while rule [B] combines quantifiers Σ and Π only with predicates, rule [B*] can combine quantifiers ∃ and ∀ with any "pseudopredicate". This has two consequences. First, as we shall see, it makes our formal language much more flexible and allows us to adequately analyze some natural language statements that have hitherto been outside the scope of our model. Second, and this is the price we have to pay for this greater flexibility, it causes our thus modified formal language to cease to be "isomorphic" to natural language and to some extent to acquire its own, autonomous syntactic structure so that the relationship of its statements to natural language sentences ceases to be a matter of direct correspondence of syntactic structures; it instead must be determined on a case-by-case basis.

So let's illustrate the newfound flexibility of our language. Sentences that have such a structure as (24) or (25) are probably not very common in natural language; i.e. the sentences of the structure *Everything is P* or *Something is P*. Sentences of the forms *Some P is Q, Every P is Q,* etc. are more common.[28] Then (26) can be analyzed as (26') and (27) as (27').

(26) *Some P is Q*

(26') $\exists x(P(x) \land Q(x))$

(27) *Every P is Q*

(27') $\forall x(P(x) \to Q(x))$

Thus, statement (28) can be captured as (28') and statement (29) as (29').

(28) *Some Italians are writers*

(28') $\exists x(\underline{Italian}(x) \land \underline{writer}(x))$

(29) *Every human is mortal*

(29') $\forall x(\underline{human}(x) \to \underline{mortal}(x))$

Now consider one more addition to our model, namely the binary predicate =. (Here we get ahead of ourselves, for in our model so far we have no binary predicates. But a binary predicate, i.e. a predicate that connects two terms into a statement, is not too problematic.) Let = function so that if $T$ and $T'$ are terms, then $T = T'$ is true iff $T$ and $T'$ denote the same thing, that is, iff $\|T\| = \|T'\|$. Given this, we can capture even more complicated syntactic constructions, such as certain noun phrases,

---

[28] However, sometimes statements of the type *Something is P* are used in the sense of *Some Q is P* – the limitation of the range of quantification to $Q$ is given by the context. For example, if I write in some mathematical treatise that there is some $x$ that is even, I probably mean that there is an even *number* $x$.

expressed in languages such as English or German using the definite article. An example of such a phrase is *the king of France* in sentence (30).

(30) *The king of France is bald*

Like sentence (28), this sentence states that sets of French kings and bald men have a non-empty intersection; however, this sentence also implies that there is exactly one king of France, so that the set of kings of France has exactly one element. The fact that a set of kings of France has exactly one element can be written in such a way that there exists some $x$ such that every $y$ that is king of France is equal to $x$; the corresponding statement is thus (30').

(30') $\exists x(\underline{PF}(x) \land \underline{wise}(x) \land \forall y(\underline{PF}(y) \rightarrow (y = x)))$

More generally, if N is a definite noun phrase and P is a predicate, then the sentence of the form (31) is reconstructed as a statement of the form (31'):

(31) *The N is P*

(31') $\exists x(N(x) \land P(x) \land \forall y(N(y) \rightarrow (y=x)))$.

This is the celebrated analysis of sentences with definite descriptions resulting from Russell (1905). Given this, it may seem that a great deal of natural language subjects are to be treated as quantifiers, indeed that perhaps the only genuine terms are proper names.

## 3.8 Predicates of greater arities

Having Fregean quantifiers on board, we can do one more extension of our model: we can add predicates of greater arities. (We have already anticipated this when adding the binary predicate =; now we officially introduce the category of binary predicates.) Consider the example (note that the examples we use are not always true):

(32) *Eco admires Schwarzenegger*

To this point, we would have to analyze it so that we would take *admires Schwarzenegger* as an unbreakable unary predicate; now we can take *admires* as a binary predicate connecting two terms

(32') *admires* (*Eco*, *Schwarzenegger*)

Now consider if we replace the terms in (32) by quantifiers:

(33) *Everybody admires somebody*

We can analyze it as

(33') $\forall x(\exists y(\text{\underline{admires}}(x,y)))$

But there is a different way of analyzing (33), revealing the hidden ambiguity of the English sentence

(33'') $\exists y(\forall x(\text{\underline{admires}}(x,y)))$

While (33') states that everybody has their own admiree, (33'') states that there is one and the same admiree for everybody.

In the same way as we introduced binary predicates, we can introduce ternary ones and, indeed, predicates of any other arity. For example, the sentence

(34) *Eco introduced Schwarzenegger to everybody*

can be most directly schematized as

(34') $\forall x(\text{\underline{introduced}}(\text{\underline{Eco}},\text{\underline{Schwarzenegger}},x))$

The addition of predicates of greater arities requires the following modifications of the basic extensional model:

The category of *n-ary predicates* (**P$^n$**) (for every $n>0$) contains an unlimited number of simple predicates. The category of binary quantifiers (**P$^2$**) contains, over and above this, the simple expression =.

[An] If $P$ is an $n$-ary predicate and $T_1$, ..., $T_n$ are terms, then $P(T_1,...,T_n)$ is a statement.

The denotation of a *n*-ary predicate is a function from U×...×U to B, i.e. the element of [U×...×U $\Rightarrow$ B] (or a set of n-tuples of individuals). The denotation of = is the function that assigns ***Tr*** to $x$ and $y$ iff $x$ is the same object as $y$.

[An'] If $P$ is an $n$-ary predicate and $T_1$, ..., $T_n$ are terms, then $\|P(T_1,...,T_n)\| = \|P\|(\|T_1\|,...,\|T_n\|)$.

Sometimes the model is extended by the introduction of an additional categories of *functors*, which let us produce complex terms (in English *the father of* can be seen as such a unary functor which takes the name *Eco* to the nominal phrase *the father of Eco*):

The category of *n-ary* functors ($\mathbf{F^n}$) (for every $n>0$) contains an unlimited number of simple functors.

If $F$ is an $n$-ary functor and $T_1$, ..., $T_n$ are terms, then $F(T_1, ..., T_n)$ is a term.

The denotation of a $n$-ary functor is a function from $U \times ... \times U$ to $U$, i.e. the element of $[U \times ... \times U \Rightarrow U]$.

If $F$ is an $n$-ary functor and $T_1$, ..., $T_n$ are terms, then $\|F(T_1,...,T_n)\| = \|F\|(\|T_1\|,...,\|T_n\|)$.

For simplicity's sake, we will not include functors in our model.

## 3.9 Modified extensional model

After this modification of the extensional model, we need to incorporate the changes we have discussed already before, namely the replacement of the rule [B] with rule [B*]. In the syntactic part of the model, this would mean replacing the quantifiers (expressions of the category $\mathbf{Q}$) $\Sigma$ and $\Pi$ by $\exists$ and $\forall$; including an unlimited number of variables $x$, $y$, ... among the auxiliary symbols; and replacing rule [B] by the rule

[B*] If $Q$ is a quantifier, $x$ a variable and $T$ a term, and $S$ a statement, then $Q(S^{T \leftarrow x})$ is a statement. (Recall that $S^{T \leftarrow x}$ denotes the result of replacing $T$ by $x$ in $S$).

To be able to modify the semantic part as needed, let us introduce the following notation convention: let the symbol $\|A\|_{\|B\|=b}$ have the denotation that the expression $A$ would have if the expression $B$ had the denotation $b$ (while preserving everything else). Thus, while

$\|\underline{writer}(\underline{Eco})\|$ is ***Tr***, $\|\underline{writer}(\underline{Eco})\|_{\|\underline{Eco}\|=\text{Schwarzenegger}}$ is ***Fa***; as according to our definition, it holds that $\|\underline{writer}(\underline{Eco})\|_{\|\underline{Eco}\|=\text{Schwarzenegger}} = \|\underline{writer}(\underline{Schwarzenegger})\|$. The necessary modification of the semantic part of $L_E$ would now mean the replacement of the rule [B′]

[B*′] if $Q$ is a quantifier, $x$ a variable, $T$ a term and $S$ a statement, then $\|Qx(S^{T\leftarrow x})\|$ is the value of the function $\|Q\|$ applied to the function that assigns the truth value $\|S\|_{\|T\|=i}$ to every element $i$ of the universe U (i.e. the value of the function $\|Q\|$ applied to the set $\{i \mid \|S\|_{\|T\|=i} = \boldsymbol{Tr}\}$).

Let's illustrate the somewhat complicated rule 4.2′ with an example where $Q$ is $\forall$, $S$ is $\underline{writer}(\underline{Eco})$ and $T$ is $\underline{Eco}$; $Qx.S^{T\leftarrow x}$ is therefore $\forall x(\underline{writer}(x))$. The set of all $i$ such that $\|S\|_{\|T\|=i} = \boldsymbol{Tr}$ is in this case the set of all those individuals to which the predicate $\underline{writer}$ applies, so it is the set of all writers. Since for a set M⊆U $\|\forall\|$(M) gives the value ***Tr*** only if M = U, i.e. if M contains all the elements of the universe, the meaning of the considered statement is the truth value ***Tr*** just when the set of writers coincides with the set of all elements of the universe, i.e. when each element of the universe is a writer.

Let us call the model that we get from $L_E$ in this way, $L_E$*. We can say that the language $L_E$* corresponds to the language of standard classical logic, i.e. classical predicate calculus of the first order. Then there is, however, one more difference concerning how we treat the language. We have moved *variables* to the role of mere auxiliary symbols; and conversely, we treat propositional operators and quantifiers, which are usually understood "syncategorematically" in predicate calculus (that is, they are not assigned any own denotations, but their semantics is implicit to the rules for denotations of statements that contain them), on an equal footing with terms and predicates.

Russell and his followers have shown that even the syntactically minimal means of a formal language such as $L_E$* (and thus standard predicate logic) are sufficient to capture many syntactically complicated statements of natural language – but only if we are savvy and resourceful enough to perform the kind of "logical analysis" we presented at the end of §3.7. However, we do not want to minimize the syntactic means of the formal languages we work with here, but rather we want to make these languages sufficiently syntactically rich so that they can serve as the most straightforward models of natural language.

## 3.10 Categorial grammar

We have seen that if a statement is composed of a term and a predicate, then Frege's maneuver leads us to take the denotation of the predicate as a function that, applied to the denotation of the term, gives the denotation of the whole statement. We have seen that a similar idea is applicable in other cases: we have explained the denotation of the quantifier that creates a statement together with a predicate as a function applied to the denotation of the predicate to give the denotation of the statement, and we have done similarly for propositional operators. This may lead us to applying this idea in general; to try to look at the denotation *of every* compound expression as the value of applying the denotation of one of its parts to the denotations of the other parts.

Such an approach was emphasized especially by those who saw a close connection between linguistic semantics and logic. The basic idea of this view of language was expressed as early as 1953 by the Israeli logician and linguist Bar-Hillel (1953, p. 50): "Each sentence that is not an element is regarded as the outcome of the operation of one sub-sequence upon the remainder, which may be to its immediate right or to its immediate left or on both sides. (...) That sub-sequence which is regarded as operating upon the others will be called an OPERATOR, the others its ARGUMENTS." We can generalize the idea expressed by Bar-Hillel even further: the result of such an operation can be considered not only every sentence, but *every compound expression at all*. As a result of such generalization, we get what is called *categorial grammar* (Casadio, 1988; Morrill, 1994). However, Bar-Hillel speaks of expressions, not their meaning, and what he means is thus not entirely clear. Later, within the set-theoretical approach to semantics, the idea became clearer: the expression is the result of the operation of one part of it on the parts of others in the sense that its denotation is the value of the functional application of the denotation of one part to the denotations of the other parts.

When we discussed compositionality (Section 1.5), we said that for every syntactic rule $R$ there must exist an operation $R^*$ such that $\|R(E_1,...,E_n)\| = R^*(\|E_1\|,...,\|E_n\|)$. That is, if there is a way (i.e. a syntactic rule) to combine some expressions into a compound expression, then there is a way to combine the denotations of those expressions into the denotation of that compound expression; in other words, for each compound expression there is some way to get its denotation from the denotations of its components. It is obvious that the simpler this method is, the simpler and thus better the resulting semantic model will be. And categorial grammar is based on the assumption that the denotation of a compound expression can always be obtained from the denotations of its components in the same simple way, i.e. by applying the denotation of one of the components, which is a function, to the denotations of the remaining ones. This means that for each rule $R$ there is an $i$ such that for all expressions $E_1, ..., E_n$, for which $R(E_1,...,E_n)$ is defined, $\|R(E_1,...,E_{i-1},E_i,E_{i1},...,E_n)\| = \|E_i\|(\|E_1\|,...,\|E_{i-1}\|,\|E_{i+1}\|,...,\|E_n\|)$.

So, if we have a syntactic rule that combines expressions of categories $A_1$, ..., $A_n$ into an expression of category $A$, then we must consider the denotations of expressions of one of the categories $A_1, ..., A_n$, say $A_i$, as "functions" applicable to denotations of expressions of the remaining categories $A_1, ..., A_{i-1}, A_{i+1}, ..., A_n$. To make this visible, we will denote category $A_i$ by the index $A/A_1,...,A_{i-1},A_{i+1},...,A_n$. Therefore, if we denote the categories of statements and terms as **S** and **T** as before, then the category of unary predicates will now be denoted as **S/T**. This designation directly indicates that a predicate supplemented by a term will give a statement; the relevant syntactic rule can be expressed in a way that is optically reminiscent of fractional shortening:

**S/T** x **T** = **S**

(an expression of category **S/T** together with an expression of category **T** gives an expression of the category **S**).

Similarly, we can view negation (and more generally any propositional operator) as an expression of the category **S/S** and any propositional connective as an expression of the category **S/S,S**. The quantifiers of $L_E$ will then be expressions of the category **S/(S/T)**.

The situation is more complicated in the case of the Fregean quantifiers of the language $L_E^*$. These are combined with statements into statements, so it might seem that they should, like propositional operators, be of the category **S/S**; however, we assigned them elements of the set [[U ⇨ B] ⇨ B], as if they were of the category **S/(S/T)**. This is because the rule for quantification in the language $L_E^*$ defies capture within categorial grammar. (We will show why this is the case in Section 3.12.)

For simplicity, we can limit ourselves to categories that have a single category behind the slash (which, however, can be compound); we can do without categories of type $A/A_1,...,A_n$ for n>1. If we have the category $A/A_1,...,A_n$, then we have a rule that combines the expressions of this category with the expressions of the categories $A_1,..., A_n$; and we can decompose this rule into *n* steps (sub-rules), in each of which one of the expressions of categories $A_1,...,A_n$ is added. Expressions of category $A/A_1,...,A_n$ will thus become expressions of category $(A/A_1)/.../A_n$ (or category $(A/A_n)/.../A_1$. Logical connectives, which, as we have seen, come out as expressions of the category **S/S,S** thus become expressions of category **(S/S)/S.** (The trick in this case is that instead of seeing the logical connectives as combining with a pair of statements into a statement, we will see them as combining with a statement into something that gives a statement when combined with another statement.)

If we have a set CAT of syntactic categories, then a categorial language based on this set can be described as follows (we will denote the language by the symbol $L_C^{CAT}$).[29] We have an unlimited number of expressions of each syntactic category, where syntactic categories are given all elements of CAT plus all B/A, where A and B are syntactic categories. If *B* is an expression of the category B/A and if *A* is an expression of the category A, then *B(A)* is an expression of the category B.

As for the denotations, we assume that for each category C∈CAT we have

---

[29] For an exact definition of the language, see §8.3.

the set $D_C$ and $D_{B/A} = [D_A \Rightarrow D_B]$. ($D_C$ is called the *domain* of category C.) If $E$ is an expression of category C, then $\|E\| \in D_C$ and $\|B(A)\| = \|B\|(\|A\|)$.

From what we said above, it follows that the language $L_E$ can be seen as part of a certain categorial language with the basic categories **S** and **T**, i.e. the language $L_C^{\{S,T\}}$. In contrast, the language $L_E*$ does not fit in $L_C^{\{S,T\}}$; what is problematic in this respect, as we have already said, is rule 2.4.

## 3.11 Type theory

In the previous section, we demonstrated the idea underlying categorial grammar as arising from general considerations of grammar and its structure. However, we can arrive at a similar idea in a slightly different way because of our efforts to deal with semantic paradoxes. In this section, which should be taken as a detour, we will explore this alternative path.

Within our model as assembled so far, a simple sentence consists of a subject and a predicate. But now consider the following sentence: its subject is the predicate of sentence (9).

(35) *Being a writer is respectable*

So it seems that what can be a predicate can also be a subject; that we must therefore admit, in addition to the syntactic rule [A], also the rule [A#]

[A#] If $P$ and $P'$ are predicates, then $P'(P)$ is a statement

We can say about a predicate that it either *applies to the given subject* (i.e. gives together with it a true sentence) or *does not apply to it* (gives together with it a false statement). But now let's imagine the predicate *not apply to itself*. What truth value would the sentence (36) have?

(36) *Not applying to itself doesn't apply to itself*

If sentence (36) is true, then what this sentence says must be the case, namely that *not applying to itself* does not apply to itself, and therefore the sentence which arises by attaching this predicate to itself is false – this, however, is nothing other than the sentence (36) itself. So if (36) is true, then it is false. Conversely, if (36) is false, then *not applying to itself* is true of itself, and (36) is true. Formally, if we have the predicate *P'* such that, for each predicate *P*, *P'(P)* holds just when ¬*P(P)* holds, then if we take *P'* as *P,* we have *P'(P')* just when ¬*P'(P')*. One way we can avoid such a paradox, the one suggested by Bertrand Russell, is to rule out the possibility of applying a predicate to itself. This means that rule [A#], as it is, cannot be accepted.

However, if we do not want to get rid of the natural analysis of sentences of type (35), we cannot completely rule out the possibility of applying a predicate to a predicate. A possible solution is to consider predicates that can be applied to predicates as predicates of a different category than those that can be applied to terms; we will therefore introduce, in addition to category **P** and rule [A], category **P²** ("second-order" predicates) and rule [A2]. Predicates like *to be respectable* can then be considered as expressions of category **P²**, i.e. as predicates of the "second-order".[30]

[A2] If *P* is an expression of the category **P** and *P'* is an expression of the category **P²**, then *P'(P)* is an expression of the category **S**.

But then we can also consider predicates that are applicable to second-order predicates and so on; in general, we can accept categories **Pⁱ** and syntactic rules [Ai] for every i.

[An] If *P* is an expression of the category **Pⁿ** and *P'* is an expression of the category **Pⁿ⁺¹**, then *P'(P)* is an expression of the category **S**.

We have concluded that a first-order predicate, which applies to a term denoting an object from U to yield a statement denoting an object from B, must denote an object from [U ⇨ B]. By analogical reasoning, we conclude that the denotation of a second-order predicate, as it applies to a first-order one, which denotes an object from [U ⇨ B] to yield a statement, which denotes an object from B, must denote an object from

---

[30] However, such a concept is not unproblematic: it assumes that the predicate as *respectable* is applicable *only* to predicates, but in reality this predicate can certainly be meaningfully applied also to a term.

$[[U \Rightarrow B] \Rightarrow B]$. For example,

$\|respectable\| =$

$\quad \|writer\| \longrightarrow Tr$

$\quad \|actor\| \longrightarrow Tr$

$\quad \|thief\| \longrightarrow Fa$

$\quad ...$

More generally, the denotation of an n-th order predicate will be a function that will assign truth values to the denotations of n-1 order predicates.

Russell actually arrived at such a language at the beginning of twentieth century; he spoke of it as a *type theory*. Russell's proposals were then elaborated in various ways; Alonzo Church (1940) combined the generalization of Russell's idea (which may further lead to categorial grammar, as outlined in the previous section) with his idea of lambda abstraction into a language, which later became the basis of modern logical systems used for natural language analysis. Let us now explain what lambda abstraction is.

## 3.12 Lambda abstraction

We have seen that we can create a matrix from an expression by "making a hole" in it, i.e. by replacing some part of it with a variable. If we fill this hole in different ways, we get different results. We can understand such a matrix as "indicating" a certain assignment or function: for example, the matrix *writer*($x$) can be understood as an indication of the function

$\quad Eco \longrightarrow writer(Eco)$

$\quad Schwarzenegger \longrightarrow writer(Schwarzenegger)$

$\quad morning\ star \longrightarrow writer(morning\ star)$

$\quad ...$

Such a function assigns expressions to expressions (so it is at the syntax level); in parallel, we can also consider the corresponding function assigning denotations to denotations (at the level of semantics), i.e. in our case the function

$\|Eco\| \longrightarrow \|writer(Eco)\|$

$\|Schwarzenegger\| \longrightarrow \|writer(Schwarzenegger)\|$

$\|morning\ star\| \longrightarrow \|writer(morning\ star)\|$

...

that is, actually the function

Eco $\longrightarrow$ ***Tr***

Schwarzenegger $\longrightarrow$ ***Fa***

Venus $\longrightarrow$ ***Fa***

...

The idea of lambda abstraction is based on introducing a new expression, of the form λ*x*(*writer*(*x*)), whose denotation is by definition just this function. More generally, if $A$ is a matrix, $\|\lambda xA\|$ is the function $f$ such that $f(\|B\|) = \|A^{x \leftarrow B}\|$ (recall that the symbol $A^{x \leftarrow B}$ denotes the variant of the expression $A$ in which the variable $x$ is replaced by the expression $B$). This means that $\|(\lambda xA)(B)\| = \|A^{x \leftarrow B}\|$, hence that the expressions (λ*xA*)(*B*) and $A^{x \leftarrow B}$ are equivalent (in the sense that they have the same denotation) and the first, more complex one can be replaced by the simpler one in any compound expression. The rule of replacing the more complex expression (λ*xA*)(*B*) with the simpler $A^{x \leftarrow B}$ is called *the lambda-conversion rule*. According to this rule, we can, for example, convert λ*x*(*writer*(*x*))(*Eco*) to *writer*(*Eco*) (because $\|(\lambda x(writer(x))(Eco)\| = \|writer(Eco)\|$).

Let's emphasize the essential difference between the terms *writer*(*x*) and λ*x*.*writer*(*x*), which, as we have just seen, both relate to the above function assigning ***Tr*** to writers and ***Fa*** to the other elements of the universe. The first *de facto* is no expression, we can intuitively look at it as a sentential scheme (which in itself has no denotation and would only get concrete denotation if *x* were replaced by a specific term in it), and it does not belong to the language of our model. The second is a real expression and its denotation is the function.

In connection with the language $L_E^*$, we considered only matrices that are created by replacing a *term* by a variable. But we can also replace expressions of other categories with variables: if we replace the predicate *writer* in *writer*(*Eco*) by the variable *p* (to replace expressions of different categories it is appropriate to use different types of variables), we get the matrix *p*(*Eco*) and by lambda-abstraction then the expression $\lambda p(p(\underline{Eco}))$. $\lambda p(p(\underline{Eco}))(\underline{writer})$ is then convertible to *writer*(*Eco*), according to the lambda-conversion rule. Similarly, we can create, say, the expression $\lambda p(\forall x(\underline{human}(x) \rightarrow p(x)))$, such that $\lambda p(\forall x(\underline{human}(x) \rightarrow p(x)))(\underline{mortal})$ then can be converted to $\forall x(\underline{human}(x) \rightarrow \underline{mortal}(x))$.

The categorial grammar we have presented above is based on the idea that some expressions name functions and all syntactic rules "express" the application of these functions. The problem was that our whole elementary language $L_E^*$ did not fit into such a framework. However, if we enrich the categorial grammar with lambda-abstraction, we get a language so rich that we can include not only $L_E^*$ but almost any "reasonable" formal language. Let us first realize exactly what $L_E^*$ has in addition to that which can be captured by pure categorial language.

What can rule [B*] do, in addition to what is done by [B]? Unlike [B], [B*] allows us to master the combination of the quantifier not only with a predicate, but also with a pseudopredicate – *viz.* matrix produced by "making a hole into" a statement. What is unmanageable within the categorial grammar, then, must be precisely this difference. However, it would disappear if we could produce all expressions like $\exists x(\underline{writer}(x) \wedge \underline{actor}(x))$ as a combination of a quantifier and a predicate; then $L_E^*$ could be *de facto* "embedded" into the categorial grammar. One way to achieve this is to introduce, as we have already considered in section 3.7, new rules for creating predicates: for example, a rule that would allow us to combine the predicates *writer* and *actor* with the operator $\wedge$ into a complex predicate (for example, we could introduce analogues of rules [C] and [D] for predicates in place of statements). However, if we introduce the lambda-abstraction rule, this problem is solved: we are able to combine the expressions *writer* and *actor* and ∧ (together with the variable *x*) into the propositional matrix *writer*(*x*)∧*actor*(*x*) and use lambda-abstraction to turn this matrix into a predicate.

Let us now show how lambda-abstraction significantly expands our possibilities for semantic analysis. Above, we presented the formula ∀*x*(*human*(*x*)→*mortal*(*x*)) as the analysis of the statement *Every human is mortal*. If we now replace the expression *mortal* in this formula with the variable *p* and use lambda-abstraction, we get the expression λ*p*(∀*x*(*human*(*x*)→*p*(*x*))) such that when applied to a predicate, it gives a statement in which this predicate is attributed to every human. If we further replace the predicate *human* with the variable *q* and use lambda-abstraction again, we get the expression λ*q*(λ*p*(∀*x*(*q*(*x*)→*p*(*x*)))), which can be considered as an analysis of the term *every*. Indeed, if this expression is applied to the predicate *human*, it gives a formula corresponding to the expression *every human*; if applied to the predicate *actor*, it gives a formula that corresponds to the expression *every actor*, etc. In a similar way we can arrive at the formula λ*q*(λ*p*(∃*x*(*q*(*x*)∧*p*(*x*)))), which corresponds to the expression *some* (and also to an indefinite article, e.g. English *a*); and we can perhaps also get the formula λ*q*(λ*p*(∃*x*(*q*(*x*)∧*p*(*x*)∧∀*y*.(*q*(*y*)→(*y*=*x*))))), which is the Russellian analysis of the definite article (the English *the*).[31]

Sentence (29) consists of the terms *every*, *human* and *mortal* (considering the clause *to be* only as an auxiliary term). As long as we work with the language L$_E$*, we can analyze such a sentence as ∀*x*(*human*(*x*)→*mortal*(*x*)), but we cannot reconstruct this formula as a categorial combination of formulas corresponding to parts of (29). As we have seen, the syntactic structure of our model language does not

---

[31] Such an analysis of the English members *a* and *the*, however often accepted, is a considerable simplification. Their truly adequate analysis can only be performed within some dynamic semantic model (see Chapter 6).

correspond to the syntactic structure of the modeled natural language. However, once we have the lambda abstraction, we can analyze the term *everyone* as $\lambda q(\lambda p(\forall x(q(x) \to p(x))))$ and then create the formula $\forall x(\underline{human}(x) \to \underline{mortal}(x))$ by first applying the formula $\lambda q(\lambda(p \forall x(q(x) \to p(x))))$ (of the category (S/(S/T))/(S/T)) to the predicate *human* (of the category (S/T)), getting the expression $\lambda p(\forall x(\underline{human}(x) \to p(x)))$ (of the category (S/(S/T))) and then to the term *mortal* (of the category (S/T)). Formula $\forall x(\underline{human}(x) \to \underline{mortal}(x))$ can therefore be considered a categorial combination of the expressions $\lambda q(\lambda p(\forall x(q(x) \to p(x))))$, *human* and *mortal.*

Lambda abstraction *de facto* allows us to combine the expressions of certain categories that cannot be combined within a categorial grammar. For example, it allows us to combine an expression of the category A/B with an expression of the category B/C into an expression of the category A/C; and this combination corresponds to the semantic level of function composition. For example, we cannot combine negation ¬, which is of the category **S/S**, in a pure categorial grammar with the predicate *mortal* because it is of the category **S/T**. However, with the help of lambda-abstraction, we can create a compound expression $\lambda x(\neg \underline{mortal}(x))$ (corresponding to the predicate *not mortal*), which is of the category of **S/T,** and a function that is its denotation is a composition of functions that are the denotations of its components. Now the following question may arise: if what is at stake is the enrichment of categorial grammar with new ways of combining categories, why not add these combinations directly to it, and why do it by detour via the λ–abstraction? Why not introduce new syntactic rules instead of introducng lambda-abstraction, such as one combining an expression of the category A/B with an expression of the category B/C into an expression of the category A/C? Indeed, some theoreticians have taken this path and have begun to examine in general the problem of what other types of category combinations could be added to categorial grammar, i.e. the problem for which X, Y and Z we can allow to combine expressions of the category X with expressions of the category Y into expressions of the category Z. Surprisingly, it turned out that the rules for combining categories are formally very similar to the rules for logical derivation. (For example, the counterpart of the fact that expressions of the category A/B can be combined with expressions of the category B into expressions of the category A is the fact that statement *A* can be derived from statements *B→A* and *B*.) The syntactic structure of possible extensions of categorial grammar can thus be shown to study the means of logic. In this context, we talk about the so-called Lambek calculus) or the logic of categories.[32]

## 3.13 Lambda-categorial grammar

If we have a set CAT of default grammatical categories, then the lambda-categorial language based on this set is a language that meets the conditions for $L_C^{CAT}$ extended by the following clauses (we will denote such a language by the symbol $L_{\lambda C}^{CAT}$)[33]:

If *B is* an expression of category B, *A* is an expression of category A and *x* is a variable of category A, $\lambda x(B^{A \leftarrow x})$ is an expression of category B/A. $\|\lambda x(B^{A \leftarrow x})\|$ is the function *f* from $D_A$ to $D_B$ such that if $d \in D_A$, then $f(d) = \|B\|_{\|A\|=d}$.

Let's demonstrate how these rules work on an example. If *B* is $\forall x(human(x) \rightarrow mortal(x))$, *A* is *mortal* and *x* is *p*, then $\lambda x(B^{A \leftarrow x})$ is the expression $\lambda p(\forall x(human(x) \rightarrow p(x)))$. According to 4.2, $\|\lambda p(\forall x(human(x) \rightarrow p(x)))\|$ is such a unction *f* that for every function *g* from [U ⇨ B] (i.e. for each set of individuals) it holds that $f(g) = \|\forall x(human(x) \rightarrow mortal(x))\|_{\|mortal\|=g}$. That is, the denotation of the expression $\lambda p(\forall x(human(x) \rightarrow p(x)))$ applied to the predicate *writer* is what would be the denotation of the statement $\forall x(human(x) \rightarrow mortal(x))$, if the predicate *mortal* denoted what the predicate *writer* actually denotes – that is, what is denoted by the statement $\forall x(human(x) \rightarrow writer(x))$.

The language $L_{\lambda C}^{\{S,T\}}$, which we will now present, is a special case of the language $L_{\lambda C}^{CAT}$ for CAT = {**S, T**}; this language is at the same time the

---

[32] See Morrill (1994).

[33] For an exact definition of the language, see §8.4.

result of extending our language $L_C^{\{S,T\}}$ (from section 3.10) by lambda-abstraction:

We have an unlimited number of expressions of each syntactic category, where syntactic categories are given as follows: **T** and **S** are syntactic categories; whenever A, B are syntactic categories, B/A is also a syntactic category. If *B* is an expression of a category B/A and if *A* is an expression of a category A, then *B(A)* is an expression of a category B. If *B is* an expression of a category B, *A* is an expression of a category A and *x* is a variable of category A, then $\lambda x(B^{A \leftarrow x})$ is an expression of category B/A.

For each category C we have a domain $D_C$ such that if *E* is an expression of the category C, then $\|E\| \in D_C$; where $D_S = B$; $D_T = U$ (where U is a given universe); $D_{B/A} = [D_A \Rightarrow D_B]$. $\|B(A)\| = \|B\|(\|A\|)$; $\|\lambda x(B^{A \leftarrow x})\|$ is such function *f* that $f(i) = \|B\|_{\|A\| = i}$.

This language essentially corresponds to the language of simple type theory, as proposed by Alonzo Church (1940) and provided with the semantics by John Kemeny (1956a; 1956b). However, the differences are in the symbolism that Church uses. He indicates categories in Greek letters; the category of terms by the letter ι and the category of statements by **o**. If α and ß are two categories, Church writes (ßα) instead of our ß/α. If *B is* an expression of category (ßα) (i.e. in our notation ß/α) and *A is* an expression of category α, then instead of our *B(A)* he writes [*BA*].

The language $L_E^*$ can now be embedded into the language $L_{\lambda C}^{\{S,T\}}$. We have seen that when predicates are viewed as expressions of the category **S/T,** propositional operators as expressions of the category **S/S**, propositional connectives as expressions of the category **(S/S)/S** and quantifiers of the category **S/(S/T),** then all $L_E^*$ rules, except rule 2.4, automatically switch to categorial syntactic rules. To capture 2.4 as a lambda-categorial syntactic rule, we only need to look at the expression $Qx(S^{T \leftarrow x})$ as an abbreviation of the expression $Q(\lambda x(S^{T \leftarrow x}))$ and at the same time stop considering the quantifier *Q* as a Fregean quantifier – that is, the expression $\exists x(\underline{writer}(x))$ will, for example, be seen as an abbreviation of $\Sigma(\lambda x(\underline{writer}(x)))$. Thus, the rule of Fregean quantification becomes a combination of the rule of "making" a predicate from a statement using lambda-abstraction with the rule of categorial combination of a quantifier with a predicate.

## 3.14 Forging functions

We have seen that set theory became such a suitable framework for doing semantics especially because it can easily accommodate the concept of function. This allowed us to make our semantic theory nicely compositional: In categorial grammar, the denotation of a complex expression was always the function value of the application of one of its parts to those of the remaining parts. Therefore, the framework of categorial grammar could make do with a single rule for composing meanings.

However, we have also seen that this was not enough. Note that everything we would have liked to have within our framework for doing semantics could be offered by the framework of categorical grammar. In particular, the rules for Fregean quantification could not be embedded into this framework. That was because this rule involved not only function application but also, in a sense, a production of functions.

To make this clearer, let us return to the analysis of definite descriptions as put forward by Russell:

(30′) $\exists x(\underline{KF}(x) \wedge \underline{bald}(x) \wedge \forall y(\underline{KF}(y) \rightarrow (y = x)))$

Presenting this formula, Russell argued that his sentence does not ascribe a property of being bald to the individual who is the king of France, but rather that it expresses something more complex which is captured by the formula – the logical form of the sentence. The formula, thus, was supposed to provide insight into the semantics of the sentence.

If we accept this analysis, then we can agree that it lets us learn something about semantics. But it doesn't automatically yield us the meaning (or denotation) of parts of the sentence; of, for example, the definite article. The article gets resolved into the logical machinery present in the logical form, but we cannot say what its meaning according to the Russellian analysis actually is.

Now enter lambda abstraction. With its help we can abstract away *bald*, getting the equivalent of the subject *KF*:

(30'a) $\lambda q(\exists x(\underline{KF}(x) \land q(x) \land \forall y(\underline{KF}(y) \rightarrow (y=x))))$

Applying lambda abstractions once more we can abstract from *KF*, getting the analysis of the definite article

(30'b) $\lambda p(\lambda q(\exists x(p(x) \land q(x) \land \forall y(p(y) \rightarrow (y=x)))))$

It is easy to see that this amounts to precisely what we previously saw as the denotation of the definite article.

The moral to be drawn from this is that having the rule of lambda abstraction we can not only compose complex denotations out of simple ones, but also decompose the complex denotations into their simper parts getting the denotation of words from those of sentences. And we not only equip the majority of expressions with functions as their denotations (which provides for the neat composition of denotations) but also provide for the decomposition of meanings of complex expressions so that we obtain these functions as denotations of their parts. In this way, semantics becomes like a huge jigsaw puzzle that can be assembled and disassembled, and the pieces of which fit nicely together when assembled properly.

## 3.15 Generalized quantifiers

At the end of this chapter, let's mention one more possible extension of our basic extensional model, an extension that is related to what is referred to as the *theory of generalized quantifiers*. We found that the basic cases of Fregean quantification are statements of the form $\exists x(P(x))$ and $\forall x(P(x))$, while the most common cases of quantified sentences in natural language are sentences of the form *every P is Q* or *some P is Q*. In the first case we can talk about *absolute quantification* (applicable to the whole universe), in the second case *relative quantification* (applying only to a limited subset of the universe). We have seen that sentences with relative quantifiers can be captured by constructions of absolute quantifiers and propositional operators (as suggested by Russell); and we have also seen that we can use lambda-abstraction to equip relative quantifiers with their semantics. However, the question arises as to whether it would not be more sensible to model the semantic of natural language by building directly on relative quantification than as a kind of "superstructure" of absolute quantification.

If we have a sentence of the form *every P is Q,* then what behaves as an absolute quantifier in that sentence is the whole phrase *every P*. We have seen that absolute quantifiers denote sets of sets of individuals; $\|every\ P\|$ will then, as we can easily see, be the set of all those sets that contain the set $\|P\|$. We will use the term *quantifier,* as before, for absolute quantifiers; we will talk about relative quantifiers, i.e. expressions like *every* and *some*, but also *most, less than ten*, etc. as *determiners.*

Thus, a common simple natural language sentence can be understood as a combination of a quantifier, which consists of a determiner and a predicate, with a predicate; that is, as a combination of a determiner and two predicates. For sentences of this type, we can formulate the following thesis: *A sentence is true if the denotation of the predicate being part of its subject and the denotation of its main predicate are related as determined by its determiner.* This hypothesis is the basic idea of the so-called *generalized quantifier theory* (Barwise & Cooper, 1981; Löbner, 1987). Within this theory, the study of the truth of basic types of sentences (and more generally also the extension of some other types of expressions) is transferred to the study of relations between pairs of sets.

Thus, what acts as a quantifier in natural language is often a determined noun phrase; it consists of some determiner (*every, some, the, a*, etc.) and a common noun phrase, which we can understand as a predicate. This leads to the syntactic rule (E).

[E] If $D$ is a determiner and $P$ is a predicate, then $D(P)$ is a quantifier

If we want to understand this rule as a rule of categorial grammar, we find determiners as expressions of category **(S/(S/T))/(S/T)**: they are combined with expressions of category **S/T** (predicates) in expressions of category **S/(S/T)** (quantifiers). The denotation of the determiner would then be a function that assigns a set of sets of individuals to a set of individuals. However, we can also understand this function as a function that assigns a truth value to a pair of sets of individuals, that is, as a set of pairs of sets of individuals, or as a relation between sets of individuals. The determiner *every* would then be the relation of inclusion, which is to say the function that applied to the pair of sets M and N gives *Tr* just when M⊆N. Similarly, the determiner *a* would be the relation of non-emptiness of intersection, that is, the function that applied to a pair of sets M and N gives *Tr,* just when M∩N≠∅. The determiner *the* would then be the function that applied to the pair of sets M and N gives *Tr,* just when M has exactly one element and this element is also an element of N.

Problems of this kind were to some extent the subject of *syllogistics,* logical teachings developed in ancient Greece, especially by Aristotle.[34] He considered the following types of sentences:

(a) *Every human is mortal*

(b) *Some human is mortal*

(c) *Some human is not mortal*

(d) *No human is mortal.*

Take, for example, (a). The term *every human* here is not the name of any one individual, it is a means of expressing something concerning all people. According to the previous section, if we consider the denotations of the terms *human* and *(being) mortal* to be subsets of the universe (namely the set of all people and the set of all mortals), then sentence (a) is true just when the first of these sets is part of the second. This means nothing more than that each element of the first set, that is, of the set of people, is at the same time an element of the second, the set of mortals; in other words, every human being is mortal.

More generally, we can easily see that each of the sentences (a) – (d) expresses some simple relationship between the respective sets (that is, the set of people and the set of mortals): we can say that (a) is true just when the set of people is part of the set of mortals; (b) is true just when

---

[34] See Gabbay and Woods (2004).

these two sets have a nonempty intersection; (c) is true if the set of people is not part of the set of mortals (i.e. if there is at least one element of the first set that is not part of the second); and (d) is true if the intersection of the two sets is empty. If we portray the set of people as a black circle and the set of mortals as a white circle, then we can say that (a) is true just when A occurs, (b) is true just when A or B occurs, (c) is true just when B or C occurs, and (d) is true just when C occurs.

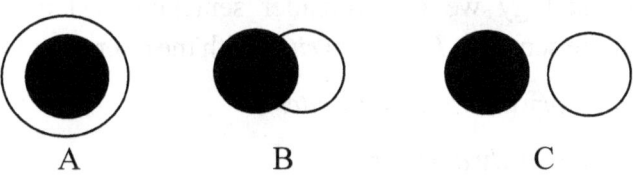

However, we can also consider more complex determiners, such as *(just) two, at most two, at least two, most, finite number*, etc.[35] Some of them are not difficult to capture: for example, the determiner *just two* corresponds to such a relationship between sets, which applied to a pair of sets A and B gives *Tr* just when A∩B has exactly two elements. This can be captured, for example, as a lambda-categorial extension $L_{E^*}$

$$\lambda p(\lambda q(\exists x \exists y(x \neq y) \wedge p(x) \wedge q(x) \wedge p(y) \wedge q(y) \wedge$$
$$\forall z.((p(z) \wedge q(z)) \rightarrow (z=x \vee z=y))))).$$

Thus, we can basically analyze sentences with this determiner within standard logic. However, there are major problems with other determiners: for example, the determiner *most* is no longer so easily reconstructed. One of the subjects of the theory of generalized quantifiers is the classification of determiners and quantifiers according to their "power", i.e. according to how "complex" a formal language is needed to capture them.[36]

---

[35] See Barwise & Cooper (1981); Löbner (1987); Peters & Westerståhl (2006).
[36] See van Benthem (1984).

# 4 Intensional model of meaning: possible worlds

## 4.1 Limits of extensional semantics

In the previous chapter, we saw that it was reasonable to consider the phrase *It is not the case that Eco is a writer* or *Eco is not a writer* as a combination of the sentence *Eco is a writer* with the negation operator. It seems that by analogy we can consider sentence (37) or (38) as a combination of the sentence *Eco is a writer* with the operator of necessity.

(37) *It is necessary that Eco is a writer*

(38) *Eco is necessarily a writer*

Just as we schematized the negation operator with the symbol ¬, we can now schematize the necessity operator with the symbol □, which can be seen as a propositional operator, i.e. an expression that applies to a statement to produce a statement. Hence we have (39) as the analysis of (37) and (38).

(39) □writer(Eco)

However, we run into unexpected difficulties when we try to determine the denotation of the symbol □ in the same way as we did in the case of ¬. As $\|\Box \underline{writer(Eco)}\|$ is certainly $Fa$ (Eco is not necessarily a writer, in any reasonable sense),

$$\|\Box\|(\|\underline{writer(Eco)}\|) = \|\Box \underline{writer(Eco)}\| = Fa,$$

and as

$$\|\underline{writer(Eco)}\| = Tr,$$

we can conclude that $\|\Box\|$ is a function that assigns the value of $Fa$ to the value of $Tr$:

$$\|\Box\|(Tr) = Fa.$$

But let's take the sentence

(40) *Eco is either a writer or not a writer*

and its version with *necessary*

(41) *Eco necessarily is either a writer or not a writer*,

i.e. $\|\Box(writer(Eco) \vee \neg writer(Eco))\|$, which is certainly true:

$$\|\Box(writer(Eco) \vee \neg writer(Eco))\| =$$
$$\|\Box\|(\|writer(Eco) \vee \neg writer(Eco)\|) = \mathbf{Tr},$$

and because it is the case that

$$\|writer(Eco) \vee \neg writer(Eco)\| = \mathbf{Tr},$$

we have

$$\|\Box\|(\|writer(Eco) \vee \neg writer(Eco)\|) = \|\Box\|(\mathbf{Tr}),$$

and therefore

$$\|\Box\|(\mathbf{Tr}) = \mathbf{Tr}.$$

However, this is obviously in direct conflict with what we arrived at above, namely that $\|\Box\|(\mathbf{Tr}) = \mathbf{Fa}.$ The reason why the meaning of the operator $\Box$ defies capture in an analogous way in which we captured the meaning of $\neg$ is that it is simply not possible to capture this meaning within the framework of extensional semantics. Let's explain why.

We have seen that the principle of compositionality entails intersubstitutivity of synonymous expressions: if the expression $E_1$ is part of the expression $E$ and if the expression $E_2$ has the same denotation as $E_1$, then we can replace $E_1$ by $E_2$ in $E$ without changing the denotation of $E$: $\|E\| = \|E^{E_1 \leftarrow E_2}\|$. One of the principles on which extensional semantics is based is the identification of the meaning of a sentence with its truth value; and hence within extensional semantics, as a special case of the principle of intersubstitutivity of synonyms, we have the principle that if the sentence $S_1$ is part of the sentence $S$ and if the sentence $S_2$ has the same truth value as the sentence $S_1$ then we can replace $S_1$ by $S_2$ in $S$ without changing the truth value of $S$. And statements with $\Box$

fundamentally violate this principle: the truth value of the statement $\Box S$ is not uniquely determined by the truth value of $S$; a true sentence may or may not be true necessarily.

The adverb *necessarily* is only one of many cases that exceed the limits of extensionality. Another is the adverb *possibly,* usually captured by the $\Diamond$ operator. Possibility is reducible to necessity and negation, because something is possible if it is not the case that its negation is necessary, i.e. $\Diamond S$ just when $\neg\Box\neg S$. Another case is a contrafactual conditional, that is, a pair of sentences connected by a connective such as *if it were the case that ..., it would be the case that ...* . (The simple implicative combination of the statements $S_1$ and $S_2$, which we schematized in the previous chapter as $S_1 \to S_2$, only says that if the statement $S_1$ is currently true, $S_2$ is also currently true, and if $S_1$ is not currently true, it tells us nothing about the truth of $S_2$. The contrafactual conditional, on the other hand, tells us that *whenever* $S_1$ becomes true (even if it is not true right now), $S_2$ would also become true.) The second statement, unlike the first, speaks not only about what is currently the case, but also about what is not, but could be. If we denote the relevant binary operator by the symbol $\Rightarrow$, then $S_1 \Rightarrow S_2$ just when $\Box(S_1 \to S_2)$. We call the operators $\Box$, $\Diamond$ and $\Rightarrow$ *modal*; and we will speak of logic with these operators as *modal logic*.

We get another type of modal operator if we look at the tense of a statement as an operator: if we consider, for example, the statement *Eco was a writer* as the application of the "past tense operator" to the statement *Eco is a writer*. Tense operators also lead us beyond the boundaries of extensionality. Consider, for example, the statement (42)

(42) *The temperature is rising.*

In the context of extensional semantics, we would say that this sentence is true just when the object temperature belongs to the set of rising objects. The sentence *Temperature is* 30°, then, would be true according to the standard extensional analysis if and only if the object temperature is identical with the object 30°. However, the truth of both sentences would then mean that the object 30° belongs to the set of rising objects; i.e. that 30° rises. This is obviously absurd.

Another classic case is the statement (43).

(43) *John seeks a unicorn*

This sentence is true in extensional semantics just when the individual John is in a certain relation to some other individual who is a unicorn; we know however very well that John can seek a unicorn quite well without there being such an object.

## 4.2  Modal logic and the concept of possible world

Aristotle had already dealt with the problems of modal logic; the foundations of the modern formalization of modal logic were laid especially by Saul Kripke (1963a; 1963b). In the 1960s, Kripke proposed a way to build simple semantics for operators like $\Box$; however, this semantics does not fit into our extensional framework. Kripke showed that we can get such a semantics if we consider the denotation of a statement not to be a truth value but a subset of a certain set; and Kripke speaks of such a set as a set of *possible worlds* – the denotation of a statement is therefore considered to be a certain subset of the set of all possible worlds. Instead of a subset of a set of possible worlds, however, we can equivalently speak of a function from possible worlds to truth values (recall that we identify a subset of a given set with the function that assigns a value of $Tr$ to all elements of that subset). If then the denotation of the statement $S$ is the function $f$, we can express that $f(w) = Tr$ by saying that $S$ is true in the world[37] $w$.

Let me emphasize that the conclusion that the denotation of a sentence is a set of possible worlds (or the corresponding function) can be arrived at by two diametrically different considerations. One way is to see possible worlds as mere auxiliary tools for the explication of certain features of natural language (more specifically modal statements and their semantic properties). We engage them on the basis of finding that they are a suitable tool for such explication, that their use leads to a relatively simple and perspicuous model that may supersede the extensional model. The other

---

[37] Or better: *about* the world – see Mates (1968).

way is to ponder the nature of statements, meaning, worlds, truths, etc. and to conclude that there are many other possible (though not actual) worlds besides our current one, and that the meaning of a statement is determined by which of these worlds this statement applies to. This is what we may call "speculative metaphysics".

Here, we prefer an approach based on the first of these paths – we try to avoid "metaphysics". The belief in the semantic usefulness of possible worlds underlying this approach stems from the fact that Kripke and others have shown that a suitable semantics for modal elements of language can be built by identifying the denotations of statements with subsets of a given set; and we have begun to call the elements of such a set possible worlds. But we could just as well call them *truth indices*, for example – they are what the truth of (empirical) statements is relative to. While within the metaphysical approach we can ask questions about the nature of possible worlds, their structure, or even their whereabouts, within our conception such questions would be beside the point. "What is a possible world?" asks American logician and philosopher of language Robert Stalnaker (1986, p. 117), and he immediately replies: "It is not a particular kind of thing or place; it is what truth is relative to, what it is the point of intellectual activities such as deliberation, communicating, and inquiry to distinguish between."

Thus, according to Kripke, the denotation of a statement is the set of possible worlds or, which we consider to be the same, the function from the possible worlds into truth values. If we denote the set of possible worlds as W, then it is an element of [W ⇨ B]. If this is the case, then, the denotation of the operator □ is a function that assigns a set of possible worlds to a set of possible worlds, i.e. the element of [[W ⇨ B] ⇨ [W ⇨ B]]. Kripke showed that in the simplest case the denotation of □ is a function that assigns the same set to the set of all possible worlds and the empty set to every other set. Therefore, if ‖□‖ is the denotation of □ within the simplest modal logic, and if M denotes any set of possible worlds, it holds

(T$\Box$) $\|\Box\|$ =

$\quad$ M $\longrightarrow$ W if M = W

$\quad$ M $\longrightarrow$ $\emptyset$ if M is a proper subset of W.

That is, the statement $\Box S$ is true (in every world), just when $S$ is true in every world; and is false (again in every world) in all other cases (i.e. whenever there is at least one world in which $S$ is false). It follows, among other things, that such a statement (and more generally any modal statement) is either true in every possible world or false in every possible world; it cannot happen to be true in one world and false in another.

Kripke and his followers dealt with many different variants of modal logic, which was due to the fact that our intuitions about modalities are not unambiguous. For example, it is clear that if something is necessary, then it is true (leading to the axiom $\Box S \rightarrow S$, which is accepted in practically all "reasonable" modal logics[38]), but it is far from obvious that if something is possible, it is necessarily possible (which would correspond to the axiom $\Diamond S \rightarrow \Box \Diamond S$ or $\neg \Box \neg S \rightarrow \Box \neg \Box \neg S$). The semantics of the necessity operator we present corresponds to the "strongest" and semantically simplest version of modal logic that C.I. Lewis called S5[39]: the principle of its semantics is, as we can see, that a statement is defined as necessarily true in any world $w$, if it is true in all worlds. The semantics of "weaker" modal logics requires, in addition to a set of possible worlds, a binary relation between them, the so-called *accessibility relation*, and a statement is defined as necessarily true in world $w$ if it is true in all worlds *accessible from w*.[40]

Take, for example, the meaning of statement (39). Returning to the

---

[38] In recent decades, modal logics have proliferated, so that now we have an almost unrestricted number of them. Not all of them, however, are meant to capture necessity, or capture it in the way we do.

[39] See Lewis & Langford (1932).

[40] See Kripe (1963b); Chellas (1980); Hughes & Cresswell (1984).

simplest modal logic we have

$$\|\Box\underline{writer}(\underline{Eco})\| =$$
$$\|\Box\|(\|\underline{writer}(\underline{Eco})\|) =$$
$$\|\Box\|(M),$$

where M is the set of all the possible worlds for which $\|\underline{writer}(\underline{Eco})\|$ has the value **Tr**. But M certainly does not contain *every* possible world, that is

$$M \neq W$$

and therefore, according to the definition of the operator $\Box$,

$$\|\Box\|(M) = \emptyset.$$

Statement (39) is therefore false (because Eco is not necessarily – i.e. in every world – a writer).

The denotations of the operators ◊ and $\Rightarrow$, which we mentioned in the previous section, can also be defined as follows:

(T◊) $\|\Diamond\| =$

$\quad M \longrightarrow \emptyset$ if $M = \emptyset$

$\quad M \longrightarrow W$ if $M \neq \emptyset$

(T$\Rightarrow$) $\|\Rightarrow\| =$

$\quad M, N \longrightarrow W$ if the set M is a subset of N

$\quad M, N \longrightarrow \emptyset$ otherwise

If we want to incorporate the new conception of meaning, based on the concept of possible world, into our previous categorial approach to language, it also means a reassessment even of our approach to logical operators and other expressions that we have managed satisfactorily within extensional semantics. For example, the denotation of the logical operator ¬ will no longer be an element of [B ⇨ B]; it will have to have the denotation from [[W ⇨ B] ⇨ [W ⇨ B]] – because ¬ combines with a statement into a statement, and the denotation of a statement is now an element of [W ⇨ B]. If we have previously considered the denotation as the function from B to B given by the prescription (T¬) (see §3.5), we will now have to consider it as a function from [W ⇨ B] to [W ⇨ B]. However, the new denotation of ¬ is not difficult to define through the old one: if $\|\neg\|_E$ is the extension of ¬ (i.e. the denotation of ¬ in the sense of the previous chapter), its intension, i.e. its denotation in the sense we introduce in this chapter, will be such a function $\|\neg\|$ that for each function $f$ from W to B and for each $w \in W$ it will be the case that

$$(\|\neg\|(f))(w) = \|\neg\|_E(f(w)).$$

In other words, the new denotation of the negation of a given statement can be determined by using the old denotation of negation to determine the truth value of that statement for every individual possible world. The transformation of classical logical connectives is similarly trivial.

However, the change will also affect the combination of terms with predicates; whereas previously the denotation of the predicate was applied to that of the term to yield the element of B, now it must yield the element of [W ⇨ B]. If we continue to consider an element of U as the denotation of a term, it would mean that we would get an element of [U ⇨ [W ⇨ B]] as the denotation of the predicate, i.e. a function that assigns an individual a function from possible worlds to truth values (according to whether the predicate applies or does not apply to this individual in that world). It is, therefore, a function which, given the individual and the possible world, yields a truth value and can therefore be easily "transformed" into a function that assigns a function from individuals to truth values to the possible world, i.e. an element of [W ⇨ [U ⇨ B]].

## 4.3 Extensions vs. intensions

The operator □ can therefore be semantically mastered by considering the denotation of a statement not its truth value in the current world, but its truth values in all possible worlds. This can be generalized: what if we considered the denotation *of each* expression not its current extension, but its extensions in all possible worlds?

Within extensional semantics, the denotation of an expression is determined by the state of the current world. Carnap (1947) argued that if we, for example, really wanted to capture the meaning of a predicate, we must take into account not only all *current* objects but also all *potential*, imaginable objects. The set of individuals falling under the predicate *human* is identical in the current world to the set of individuals falling under the predicate *featherless biped* (at least this is usually assumed); yet these two predicates certainly have different meanings. Carnap points out that it is the merely possible objects that fall under these concepts that distinguish them from each other: we can certainly imagine a featherless biped that is not human.

One way to take this into account would simply be to include not only all current (real) individuals, but all potential individuals directly into the universe of discourse. However, this would seem to completely blur the boundary between the possible and the real, and between the actual and the imaginary. However, the example of Kripke's modal logic suggests a different path: to keep the current universe of discourse and the whole current world as it is, and to think of universes and alternative worlds next to it. To say that something is possible is *de facto* nothing more than to say that it is real in some possible world; to say that the meaning of the predicate *human* is given by the set of all possible (imaginable) people is nothing more than to say that it is given by the sets of people of all possible worlds.

Therefore, assuming that we have a set of possible worlds, we can declare the denotation of the predicate *human* to be a function from this set of possible worlds, namely a function that assigns, to every possible world, the set of people of the world. In general, we can declare the denotation of a predicate to be a function that assigns, to every possible world $w$, the extension of that predicate in $w$. The denotation of a predicate thus becomes an element of the set [W ⇨ [U ⇨ B]], in the spirit of what we said at the end of §4.2. The meaning of any expression can then be declared as a function that assigns, to every possible world, the extension of this expression in $w$. The meaning of an expression conceived in this way is called its *intension*; intension is thus *de facto* generally an extension relativized to possible worlds. The semantics that identifies meaning with intension is called *intensional* semantics. (However, the term *intensional semantics* is not always used in this well-defined sense – *intensional* is sometimes called just any semantics that is not extensional, not necessarily just semantics based on the notion of a possible world.)

The denotation of a statement within an intensional model of meaning is thus a function from possible worlds to truth values or the corresponding set of possible worlds; the meaning of a particular statement is the set of precisely the possible worlds in which this statement is true. The denotation of $\|\mathit{writer}(\mathit{Eco})\|$ is the set of all such worlds in which Eco is a writer. The denotation of a predicate is a function from possible worlds to sets of individuals; in particular, it is the function that assigns to each possible world the set of individuals to which this predicate applies in this world. For example, the denotation of $\|\mathit{writer}\|$ is a function which assigns to each possible world the set of all those individuals who are writers in this possible world. The meaning of a term is a function that assigns to each possible world an individual, namely the individual to which this term refers in this possible world (if such an individual exists there); $\|\mathit{Eco}\|$ is thus a function that assigns to a possible world an individual who is Eco in this world.[41]

## 4.4 Kripkean semantics

We saw that Kripke not only discovered that employing possible worlds lets us build a nice semantics for the simplest modal logic, but also that the engagement of the relation of accessibility lets us build semantics for various other kinds of logics.

We have seen that a lot of sentences involving "necessarily" and/or "possibly" do not appear to have any determinate truth value. Therefore, it is quite natural that we have different modal logics, none of which is excluded by our intuitions. And Kripke discovered that we can have a versatile semantics adjustable to many different logics if we

---

[41] However, proper names are sometimes taken as having no intensions, only extensions (being, as Kripke. 1972, called them, *rigid designators*).

engage, besides possible worlds, also the so-called *relation of accessibility* among them.

The relation distinguishes between those possible worlds that are to be considered as alternatives of a given world and those which are not. The former are taken to be accessible from the given world. This yields us a binary relation that for each world specifies those worlds that are accessible from it. As an example, we can consider the relation that interrelates two possible worlds if and only if the same natural laws hold in them. The resulting notion of necessity is then the physical kind of necessity, the necessity engendered by natural laws.

It turns out that certain formulas of the language of modal logic correspond to certain properties of the accessibility relation in the sense that the formulas are true in every universe of possible worlds with the accessibility relation with the corresponding property, and, conversely, if the formulas are true with respect to such a universe, the relation is bound to have the property.

Thus, consider the following formula:

(T) $\Box A \rightarrow A$.

and look at what it says from the viewpoint of semantics. As $\Box A$ says that A is true in all words accessible from any given world $w$, the whole (T) says that the truth of a formula in every possible world accessible from $w$ entails its truth in $w$. It is easy to see that this is guaranteed if and only if $w$ is always among the worlds accessible from $w$ – i.e. if the accessibility relation is reflexive. In this sense the reflexivity of the accessibility relation and the formula (T) are two sides of the same coin.

Or look at the formula

(B) $A \rightarrow \Box \Diamond A$

It is true if the truth of A in $w$ implies that every world accessible from $w$ has a world accessible from it such that A holds there. This amounts to the accessibility relation being symmetric: indeed, if it is symmetric then $w$ is always world accessible from a world accessible to it.

It follows that there is a sense in which doing modal or logic is structuring the universe of possible words and hence perhaps doing a kind of (non-speculative) metaphysics.

## 4.5 Temporal logic

Worlds evolve. This means that what held true in a world at one time may no longer hold there at a later time. Given this, it would seem that any talk about extensions in a possible world is incomplete, that we would need to say at which time of the possible world it is that we are talking about. And indeed, all the intensional logicians we are talking about are aware of this. Where we talked, for simplicity, about possible worlds they were talking about word-time pairs. In what follows, we will continue to ignore time to reach as simple of language models as is possible; but in this section we will say something about it.

The dependence of extension on time may be studied relatively independently of its dependence on the world. Indeed, in our example about raising temperature, (42), that we presented as one of the emblematic examples of intensionality, it is the dependence on time which matters. Temporal logic was in fact elaborated by Prior (1957) earlier than Kripke provided his possible worlds semantics for modal logic. And there are some parallels between these two kinds of logic: for example, there is the obvious sense in which the behavior of temporal adverbs *always* and *sometimes* w.r.t. time is similar to that of the modal adverbs *necessarily* and *possibly* w.r.t. possible worlds.

To be sure, there is the difference that time moments are linearly ordered (disregarding speculations about branching time, etc.), while for possible worlds there is no such ordering. This makes some of the logical constants of temporal logic different from those of modal logic. For example, we can have a constant $\mathbf{F}_t$ such that $\mathbf{F}_t A$ is true just in case $A$ is true in every time moment greater or equal to $t$. If now $t_0$ is the actual time ("now"), then $\mathbf{F}_{t_0} A$ can be read as "$A$ holds from now on". (But note that we could similarly mark out the actual world, as, say, $w_0$; as the set of possible worlds is unordered, there is no similar kind of use for it.)

As a* result, the combination of those two systems of logic into the system of intensional logic can be pretty complicated; and therefore we disregard these complexities and we will only consider the dependence of extension on possible words.

## 4.6 From dependency to function

It is useful to realize that the train of thought that brought possible worlds into the apparatus of a formal semantics can be seen as related to that which led to the identification of the denotations of predicates (and consequently other types of expressions) with functions. This may not be immediately obvious because here the analogous maneuver works, as it were, more behind the scenes, inside of the very apparatus of semantics. Frege's maneuver was driven by the fact that a predicate yields the truth-value of a sentence given the subject with which it is combined into the sentence. Therefore, it suggests identifying the denotation of the predicate with the function mapping denotations of subjects, i.e. individuals, on the corresponding truth values.

However, in the case of many subject-predicate sentences, the truth value of the sentence is not uniquely determined by the denotations of its subject and predicate – we need, in addition to this, the "circumstances". (The sentence *The King of France is bald* is false, or perhaps truth value-less, here and now, but it might have been true some centuries ago – here is where we face the limits of extensional semantics.) The truth value, that is to say, *depends* on the "circumstances". And to neutralize the dependency, we must incorporate the "circumstances" into the denotation of the sentence – that is, to abandon the idea that the denotation of the sentence is its truth value and replace it by the idea that it is a function from possible worlds to truth values.

The same maneuver can be carried out not only for sentences, but also for expressions of other categories (perhaps with some exceptions). At the end of Section 4.3, we saw that the denotations of predicates must be transformed from [U ⇨ B] to [W ⇨ [U ⇨ B]]. But this is not enough.

As the denotations of terms will be also transformed, from U to $[W \Rightarrow U]$, those of predicates will have to be further transformed to $[W \Rightarrow [[W \Rightarrow U] \Rightarrow B]]$. We will see, in the following sections, the degree to which the resulting semantics for intentional logic is more complicated than that of the extensional one.

There is, however, one aspect of this situation which should be pointed out. In the original version of Frege's maneuver we reached functions from individuals to truth values, which didn't seem overly problematic. The same holds for all other kinds of functions in the extensional model. Here, however, the situation becomes much more complicated. What we need to be included into the model are the "circumstances" that got transformed, through the efforts of Carnap and Kripke, into possible worlds.

We have already avoided the question as to what a possible world is, relegating it to "metaphysics". Here we will do no better; but we add a warning. The intensional model of meaning appears to be – just like the extensional one – a mathematical entity which lets us use various mathematical methods to deal with it; however, it is very good to keep in mind that in the foundation of this edifice there lies this very vague and shaky concept of a possible world that is not delimited with mathematical precision.

Anyway, the general pattern shows that if an expression appears to denote various objects depending on something, we let the denotation absorb the something and become a function. Then we can say that the expression denotes the function independently of anything.

## 4.7 Possible worlds and individuals

In extensional semantics, we worked with the universe of discourse, i.e. the set of all individuals eligible as denotations of terms. In intentional semantics we have, over and above this, the set of possible words. The question now is what is the relationship between these two sets.

Let us return to the statement

(29′) $\forall x(\underline{human}(x) \rightarrow \underline{mortal}(x))$

and let us add the necessity operator. This can be done in two ways, so we have two versions of the "necessitation" of (29′).

(29″) $\Box \forall x(\underline{human}(x) \rightarrow \underline{mortal}(x))$

(29‴) $\forall x \Box(\underline{human}(x) \rightarrow \underline{mortal}(x))$

The first version says, roughly, that it is necessary that every human is mortal. The second says that for every individual it's necessary that if it is human, then it is mortal. Are these two different messages, or one and the same?

Remember that necessity amounts to quantification over possible worlds, so the box could be replaced by another general quantifier ranging over the set of possible words. (This is true for S5; in other modal logics the operator quantifies only over those possible words which are accessible.) If this is so then $\Box\forall x$ is the same as $\forall x\Box$, and consequently (29″) is synonymous with (29‴).

But should we treat the set of individuals and the set of possible words as independent of each other? Is it not rather the case that every possible word has its own set of individuals? Can it not be that an individual is in the universe of a world without being in the universe of another one?

If we assigned its own universe of discourse to every set, then $\Box\forall x$ would amount to quantifying over all (accessible) possible words and then over the individuals of the possible worlds, while $\forall x\Box$ would have no clear sense. It seems to attempt to quantify over individuals without specifying which set of individuals it is. Perhaps we could take it as quantifying over the union of all universes of all the worlds, but this would be a little bit cumbersome.

Is there a logical relationship between (29″) and (29‴)? Is it the case that[42]

---

[42] According to R. C. Barcan (1946), these statements are instances of what is usually called Barcan formulas.

(?) $\Box\forall x(\underline{human}(x) \to \underline{mortal}(x)) \to \forall x \Box(\underline{human}(x) \to \underline{mortal}(x))$

or

(?) $\forall x \Box(\underline{human}(x) \to \underline{mortal}(x)) \to \Box \forall x(\underline{human}(x) \to \underline{mortal}(x))$?

Let us start with the first formula. Its antecedent states that $\underline{human}(x) \to \underline{mortal}(x)$ holds for every individual (as the value of $x$) of every possible world. It is not so easy to interpret the consequent. However, given that each world has its universe of individuals, its truth value in a possible world $w$ would be **Tr** iff that of $\underline{human}(x) \to \underline{mortal}(x)$ is **Tr** for every individual of $w$ in every world accessible from $w$.

What if an individual of the actual world does not exist in a possible world accessible from it? Then, obviously, $\underline{human}(x) \to \underline{mortal}(x)$ is not true for that individual in that possible world. In other words, the formula holds iff the universe of every world accessible from a given world $w$ is the superset of that of $w$.

By parity of reasoning we easily find out that the second formula holds iff the universe of every world accessible from a given world $w$ is the subset of that of $w$. Put together, both the formulas hold iff the universe of every world accessible from a given world $w$ is the subset and the superset of that of $w$, i.e. if it is equal to the universe of $w$.

So is it better to have one common universe for all worlds, or an idiosyncratic universe for every world. It would seem that what we really need is that one and the same individual be an element of different worlds? Consider a contrafactual such as *If I were the president of Russia ...* . If we are to analyze such a sentence, we need words different from the actual one in that I am the president of Russia in them, but it also important that it is the real me, the same person as in the actual world. From this viewpoint, it seems that what we need is the same universe across different possible worlds.

From the other side, it would seem that our world could contain an individual, which is utterly absent in another world (or vice versa).

This would rather speak for the own universe of every world. Hence, we would need an intermediate solution: an individual might be an element of different worlds, but not necessarily of all worlds. This solution however would be, technically, quite complicated, hence there may be a reason to stay with one of the previous ones.

## 4.8 Montague's grammar and locally intensional logic

We have so far seen the semantic analysis of language as a construction of a formal language serving as a model of natural language. Such a formal language consists of syntax, parallel (maybe imperfectly) to the syntax of natural language, and of semantics, providing the expressions of the artificial language with denotations (individuals, truth values or sets) which are to explicate (or "materialize") the semantics of natural language. Montague turns logical analysis into a two-stage process: first he reconstructs natural language by a certain formal language that is categorical in our sense of the word but has no semantics, and then "translates" that language into another formal language, a language of intensional logic that already has semantics. The first stage of such a reconstruction is usually called *Montague grammar* (MG), the second *Montague intensional logic* (MIL).

Montague grammar basically follows the categorial approach to language, as outlined in the previous chapter. However, Montague's approach also has some peculiarities that are not directly related to its intensionality. One of them is his treatment of noun phrases. In the previous chapter, we assumed that a simple sentence can be a combination of a term (expression of type **T**) and a predicate (expression of type (**S/T**)); we later admitted that it can also be a combination of a quantifier (expression of type (**S/(S/T)**)) and a predicate. We then came to the conclusion that, with the exception of statements in which the subject is a proper name, the subject should be understood as a quantifier. Montague goes one step further: he understands *every* sentence as a combination of a quantifier and a predicate, and thus he basically gives up the reconstruction of a sentence as a combination of a predicate and a term. Montague's step is based on the fact that $P(T)$ is equivalent to $\lambda p(p(T))(P)$ (the expressions have the same denotation), and that we can therefore identify the term $T$ with the quantifier $\lambda p(p(T))$. (It is basically the identification of an individual with the set of all its 'properties', *viz.* with the set of all the sets containing it.) Thus, the sentence *Eco is a writer* is reconstructed by Montague not as *writer*(*Eco*), but as *Eco*$^*$(*writer*), where *Eco*$^*$ is an expression of type (**S**/(**S**/**T**)) such that *Eco*$^*$ = $\lambda p(p(Eco))$.

Like $L_\lambda^{\{S,T\}}$ in the previous chapter, MG and MIL have two basic categories, namely the category of terms and the category of statements; the other categories are then derived from them. In addition, however, there is a category (or rather quasi-category) in the MIL that corresponds to the newly introduced domain of possible worlds and which we will denote by the letter **I**. It is not a full-fledged category, it can only appear after the slash in indexes of compound categories: if A is a category, A/**I** is also a category (but not **I**/A). However, while the expressions of the category A/B, where B is not **I**, combine with expressions of the category B into expressions of the category A, expressions of the category A/**I** are not intended to associate with expressions of the category **I** because MIL does not have any such expressions.

The transition from MG, which is an instrument of an immediate categorial reconstruction of the structure of natural language, to MIL generally works by "transforming" every type A/B of MG to the type A/(B/**I**); then every expression of the form *A*(*B*) is "translated" into the expression *A'*(^*B'*). Thus, if the sentence *Eco is a writer* at the MG level is reconstructed as *Eco*$^*$(*writer*), where *Eco*$^*$ is of the type **S**/(**S**/**T**) and *writer* of the type **S**/**T**, then at the MIL level it is reconstructed as *Eco*$^{*\prime}$(^*writer'*), where *writer'* is of type **S**/(**T**/**I**) and *Eco*$^{*\prime}$ is of type **S**/((**S**/(**T**/**I**))/**I**). However, MIL has a notation completely different from both categorial grammar and Church's type theory: **e** is used instead of **T**, **t** instead of **S** and **s** instead of **I**; instead of A/B it is then written <B,A>. The word *writer* would therefore be, in Montague's notation, of type <<s,e>,t> and the expression *Eco*$^{*\prime}$ of type <<s,<<s,e>,t>>,t>. In the interest of continuity, we will not use this notation here and will adhere to the one we have used so far.

Let us first formulate the language $L_{MIL}$ of (simplified) Montague intensional logic.[43] We have an unlimited number of simple expressions of each syntactic category where the syntactic categories are given as follows: **T** and **S** are syntactic categories; whenever A, B are syntactic categories, a syntactic category is also B/A; and whenever A is a syntactic category, a syntactic category is also A/**I**. In addition, we assume that we have auxiliary, *syncategorematial symbols*: parentheses, $\lambda$, ^, ˅ and an unlimited number of variables for each syntactic category.

As for building complex expressions out of simple ones, if $B$ is an expression of the category B/A and $A$ an expression of the category A, $B(A)$ is an expression of the category B; and if $B$ *is* an expression of the category B, $A$ an expression of the category A and $x$ a variable of category A, $\lambda x(B^{A \leftarrow x})$ is an expression of the category B/A. Moreover, if $A$ is an expression of the category A, then $^\wedge A$ *is* an expression of the category A/**I**; and if $A$ is an expression of a category A/**I**, then $^\vee A$ is an expression of the category A.

As for the denotations of expressions, we assume that for each category C we have domains $D_C$ and $S_C$ such that $S_C$ is [W ⇨ $D_C$] and if $A$ is an expression of the category C, then $\|A\| \in S_C$; where $D_T$ = U (where U is a given universe); $D_S$ = B; $D_{A/B}$ = [$D_B$ ⇨ $D_A$]; and $D_{A/I}$ = [W ⇨ $D_A$] (where W is a given set of possible worlds). The denotations of complex expressions are then computed from those of their parts according to the following prescriptions (where $\|A\|^w$ is a shortcut for $\|A\|(w)$, i.e. for the value of $\|A\|$ for the possible world $w$):

$\|B(A)\|^w = \|B\|^w(\|A\|^w)$

$\|\lambda x(B^{A \leftarrow x})\|^w$ is the function $f$ that for each $i \in D_A$, $f(i) = \|B\|_{\|A\|^w = i}$

$\|^\wedge A\|^w = \|A\|$

$\|^\vee A\|^w$ is the function $f$ that for each $w' \in W$, $f(w') = \|A\|^{w'}(w')$.

Let's illustrate what the analysis of the statement (43), i.e. the statement *John seeks a unicorn*, would look like. At the MG level, Montague reconstructs it as a combination of the quantifier <u>*John*</u>* (an expression of

---

[43] For an exact definition, see §8.5.

the type **S/(S/T)**) and the predicate *seeks a unicorn* (an expression of the type **S/T**); the expression *seeks a unicorn* is then further reconstructed as a combination of the quantifier *a unicorn* of type **S/(S/T)** with the expression *seek* of type **(S/T)/(S/(S/T))**. (The quantifier *a unicorn* is then further understood as a combination of the determiner *a* of the type **(S/(S/T))/(S/T)** and the predicate *unicorn* of the type **S/T**). At the level of intensional logic, the whole statement is further reconstructed by the formula *John*$^{*}$'(^*seek*'(^*a*'(^*unicorn*'))), where *John*$^{*}$' is of type **S/((S/(T/I))/I)**, *seek*' is of the type **(S/(T/I))/((S/((S/(T/I))/S)/S))**, *a*' is of the type **(S/((S/(T/I)))/I))/((S/(T/I)/I))** and *unicorn*' is of the type **S/(T/I)**. We can clearly see Montague's reconstruction in the following diagram:

*John seeks a unicorn*

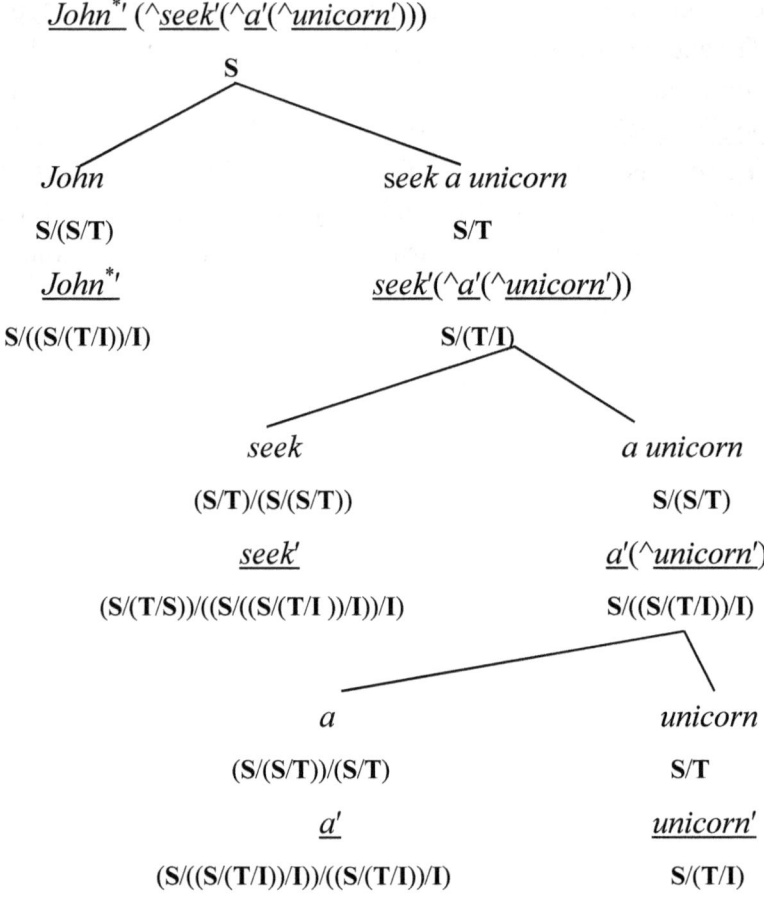

*seek'* is thus *de facto* a relation between two intensions, between the intensions of the expressions *John*$^{*\prime}$ and *a'*(*unicorn'*). It is similar with the sentence *John finds a unicorn*, which is analyzed as

$\quad$ *John*$^{*\prime}$('^'*find'*('^'*a'*('^'*unicorn'*))).

By definition, *John*$^{*\prime}$ is equal to $\lambda p(p(\char`\^ John'))$ (where *John'* is of type **T**), and

$\underline{John}^{*\prime}(^{\wedge}\underline{seek}'(^{\wedge}\underline{a}'(^{\wedge}\underline{unicorn}')))$

is therefore equivalent to

$\underline{seek}'(^{\wedge}\underline{a}'(^{\wedge}\underline{unicorn}'))(^{\wedge}\underline{John}')$.

Similarly

$\underline{John}^{*\prime}(^{\wedge}\underline{find}'(^{\wedge}\underline{a}'(^{\wedge}\underline{unicorn}')))$

is equivalent to

$\underline{find}'(^{\wedge}\underline{a}'(^{\wedge}\underline{unicorn}'))(^{\wedge}\underline{John}')$.

In the case of *find'*, Montague further introduces the so-called meaning postulates guaranteeing the existence of the expression $\underline{find}^{*\prime}$ of type **(S/T)/T** such that

$\underline{find}'(^{\wedge}\underline{a}'(^{\wedge}\underline{unicorn}'))(^{\wedge}\underline{John}')$

is equivalent to

$\underline{unicorn}'(\lambda y(\underline{find}^{*\prime}(\underline{John}')(y)))$.

and so in this case there is an extensive detour to a normal, extensional analysis. However, a similar semantic postulate is not available in the case of *seek'* – this verb is *irreducibly* intensional.

A somewhat simpler example is the analysis of statement (42), i.e. the statement *The temperature is rising*:

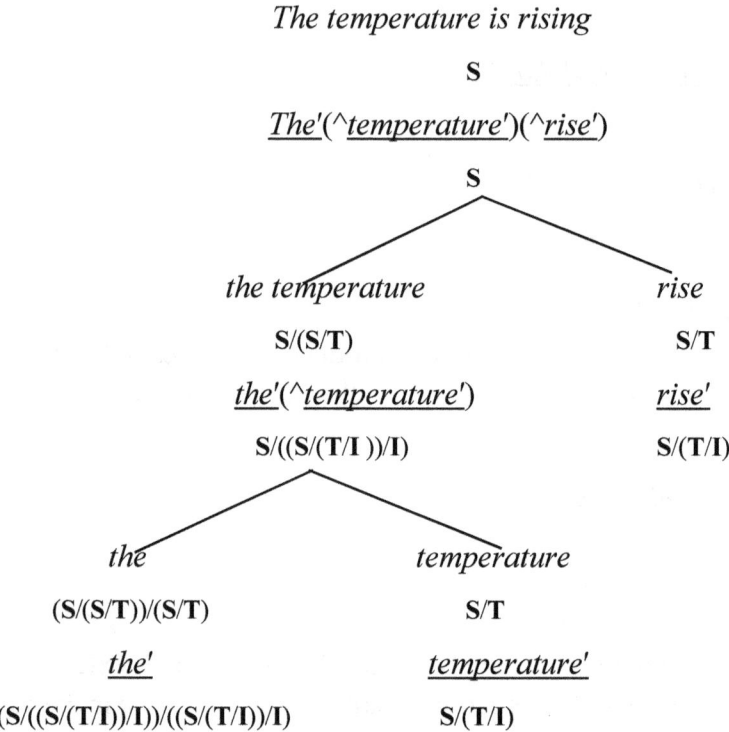

If we now use the definition of the logical operator *the'* (corresponding to the English definite article), we arrive at an analysis according to which this statement expresses the attribution of a property expressed by the expression *rise* to the intension of the expression *temperature*; while the statement *The temperature is* $30^0$ attributes the property expressed by the expression *be* (*equal to*) $30^0$ to the extension of the expression *the temperature*. (It is again, as in the case of seeking and finding a unicorn, given that the first of these properties is intensional and the second extensional; i.e. that for the second, unlike the first, there is a semantic postulate that allows it to reduce its primary intensional analysis to an extensional one.)

Thus, Montague's analysis works by first "intensionalizing" all expressions through the "translation" between MG and MIL, and then a substantial portion of them are "extensionalized" again using semantic postulates. We have just seen how complicated and intransparent the final apparatus is.

## 4.9 Montague's approach to intensions

One way to approach the new kind of semantics is to try to maintain extensional semantics wherever possible, and to resort to intensions only where extensions are not enough. We can insist that the meaning of an expression is "normally" its extension and that only in certain cases does its intension take its place. This is the idea on which Richard Montague (1974) based his intensional semantics at the turn of the 1960s and 1970s; his approach was then accepted as the standard by many logicians and linguists. If we accept Montague's conceptual framework, we must continue to consider its extension as the meaning of the term in the true sense of the word. The idea behind Montague's approach lies in the fact that what would normally be the intension of an expression is in some contexts "elevated" to its extension. In the sentence *John found a rhino*, the extension and therefore the meaning of the term *a rhino* is some actual individual; the sentence *John seeks a unicorn*, however, cannot be analyzed in this way, it must be analyzed so that the extension (and therefore the meaning) of the term *a unicorn* becomes what is normally its intension, i.e. some function from possible worlds to individuals (what is then sometimes called *an individual concept*). In this sense, Montagovian semantics remains formally extensional: meaning is always the extension.

Montague introduces the operator "^" that modifies the expression $E$ into an expression whose extension is the intension $E$, which thus "intensionalizes" it; if we denote the extension of $E$ as $\|E\|_E$ and the intension of $E$ as $\|E\|_I$, then

$$\|{^\wedge}E\|_E = \|E\|_I.$$

The operator "˅" is then complementary to "^" in the sense that

$$\|{^\vee}{^\wedge}E\|_E = \|E\|$$

More generally, if the extension of the expression $E$ is a function from possible worlds, then

$$\|{}^{\vee}E\|_I = \|E\|_E$$

In this case, the same applies

$$\|{}^{\wedge\vee}E\|_E = \|E\|_E$$

With the help of the operator $^\wedge$, Montague then analyzes the expression *seek a unicorn* not as *seek(unicorn)*, but rather as *seek(^unicorn)*; therefore, the denotation of *seek* is not a relationship between two extensions but between extension and intension. Similarly, *temperature is rising* will not be analyzed as *rise(temperature)*, but as *rise(^temperature)*.

Hence, though Montague did not put it like this, intension can be generally seen as a function that assigns, to each possible world, the extension in that possible world. We therefore assume that we have a set W of possible worlds and that $\|E\|_I$ is a function with its domain equal to this set; and that $\|E\|_E = \|E\|_I(w_a)$ for a certain $w_a \in W$, namely for that $w_a$ which corresponds to our present, actual world. We will write simply $\|E\|$ instead of $\|E\|_I$, and $\|E\|^w$ instead of $\|E\|(w)$; let us remember, however, that using such a notation which Montague himself declares to be the denotation of the expression $E$ is not its intension $\|E\|$ but its (actual) extension $\|E\|^{w_a}$.

## 4.10 Globally intensional logic

We will show the principles of the globally intensional approach with the example of a language that we will call $L_{TIL}$, which is a simplified version of Tichý's language of his *transparent intensional logic*. Note that it differs formally from $L_{MIL}$ only in the treatment of possible worlds. However, the fundamental difference between $L_{\lambda C}^{\{S,T\}}$ and $L_{TIL}$ is, as mentioned above, how these languages are related to natural language. A natural language expression that is standardly reconstructed by an expression of category A of $L_{\lambda C}^{\{S,T\}}$ is reconstructed by an expression of the category **A/I** of $L_{TIL}$. For example, while statements within $L_{\lambda C}^{\{S,T\}}$

were considered as expressions of category **S**, within $L_{TIL}$ they are considered as expressions of category **S/I**; predicates are considered not as expressions of category **S/T** but expressions of category **(S/T)/I** , etc.

The language $L_{TIL}$[44] is based on the assumption that we have an unlimited number of simple expressions of each syntactic category, where the syntactic categories are given as follows: **T, S** and **I** are syntactic categories; whenever A, B are syntactic categories, then so is B/A. In addition, we assume that we have auxiliary, *syncategorematical symbols*: parentheses, λ and an unlimited number of variables for each syntactic category.

As for the compound expressions, if *B* is an expression of category B/A and *A* is an expression of category A, then *B(A)* is an expression of category B; and if *B* is an expression of category B, *A* an expression of category A and *x* a variable of category A, then $\lambda x(B^{A \leftarrow x})$ is an expression of category B/A.

As for the denotations, for each category C we have a domain $D_C$ such that if *A* is an expression of category C, then $\|A\| \in D_C$; where $D_T = U$; $D_V = B$; $D_I = W$ (where U is a given universe and W is a given set of possible worlds); and $D_{B/A} = [D_A \Rightarrow D_B]$. Denotations of compound expressions are computed according to the following prescriptions:

$\|B(A)\| = \|B\|(\|A\|)$

$\|\lambda x(B^{A \leftarrow x})\|$ is the function *f* such that $f(i) = \|B\|_{\|A\|=i}$

To show the difference between the Montagovian approach and Tichý's approach, let's return to the example (42), i.e. the rising temperature example.[45] Tichý, like Montague, would analyze this sentence as

---

[44] For its precise definition, see §8.5.

[45] Tichý's analysis is not easy to compare with Montague's. For Montague, as we have seen, the analysis of the syntax of natural language and the explicit capture of the path from the expression to the appropriate logical formula (which leads through Montague's grammar) are also essential; Tichý, on the other hand, does not deal much with the considerations of natural language syntax and takes the path from expression to the appropriate formula as

attributing a property expressed by the predicate *rise* to the intension of the term *the temperature*. This means that according to him, *rise* would have to be analyzed as the predicate *rise'* of type **(S/(T/I))/I**, and if we analyze the term *the temperature* as *the(temperature)* of type **T/I**, it would lead to the analysis of the whole sentence as the formula $\lambda w((\underline{rise}(w))(\underline{the(temperature)}))$. In contrast, the expression *to be* $30^0$ would be analyzed as the expression $\underline{30^0}$ of type **(S/T)/I**, i.e. as a property of individuals, and the sentence *The temperature is* $30^0$ would then be analyzed as $\lambda w((\underline{30^0}(w))((\underline{the(temperature)})(w)))$. The difference is obviously that in the first case the formula simply contains the term *the(temperature)*, while in the second case it contains the expression *(the(temperature))*(w), which is the value of the intension of *the(temperature)* for the possible world *w*. Tichý says that in the first case the term *the(temperature)* is in the supposition *de dicto* (to determine the extension of the whole statement in the possible world *w* the whole intension of this expression is needed), while in the second case in the supposition *de re* (to determine the extension of the whole statement in the possible world *w* we need just its extension in *w*). Let us recall that, according to Montague, the term would first be preceded by the intensionalization operator ^, which we would be able to get rid of in the latter case with the help of the semantic postulate, while in the former we would not. Thus, if Montague needs the intensionalization operator to replace the extension of an expression with its intension in non-extensional contexts, it is enough for Tichý to apply the intension to produce the appropriate extension from the intension in intensional contexts.

## 4.11 Two-sorted type theory

It is possible to raise serious objections to the Montagovian approach to semantics, both conceptual and technical. Montague's contextualism, that is, his assumption that a term "normally" means its extension and only in special cases (in a specific set of contexts) its intension, raises principled

---

something more or less intuitively obvious.

objections. If the denotation of the expression is its extension, then in cases like *John seeks a unicorn*, the denotation of the whole sentence is not determined by the denotations of the parts: to "calculate" the extension of the whole sentence, we need not the extension but the intension of the term *a unicorn*. Montague says that the denotation of the term *a unicorn* in such a sentence is ("extraordinarily") its intension; but that means it's an ambiguous term. Montage's logic behaves as if the intension of an expression could be derived from its extension; but this is obviously problematic. Obviously, there can be (except in trivial cases) no function $f$ such that for every expression $E$, $\|{^\wedge}E\| = f(\|E\|)$), that is, a function that would assign the appropriate intension to an extension: the same extension can be clearly shared by many intensions (that is the point of the intensional approach), and therefore no extension has any single "appropriate" intension. The operator $^\wedge$ therefore does not satisfy the principle of compositionality and is thus not a *de facto* correctly defined operator in this sense.

Objections of a technical nature can be raised against the disproportionate complexity of Montague's system: the combination of the expression $B$ of the type B/A with the expression $A$ of the type A analyzes Montague in general not as $B(A)$ but as $B(^\wedge A)$, and except for "abnormal" contexts he introduces postulates that allow him to convert this connection back to the form $B(A)$. The result, as we have seen above, is quite intransparent formulas full of the operators $^\wedge$ and $^\vee$.

We would eliminate both shortcomings of Montague's "locally intensional" approach to semantics by moving to the "globally intensional" approach. Such an approach was promoted in the 1970s by the Czech logician Pavel Tichý (1971; 1978; 1988): according to his proposals, the meaning of the expression is simply its intension, and the concept of extension is not essential for semantic theory. Tichý showed that within the globally intensional approach, it is possible to define intensional semantics very simply: through the language of Church's type theory, i.e. our language $L_{\lambda C}^{\{S,T\}}$, with the addition of the third basic type, namely the type of possible worlds. However, such a language is used to reconstruct natural language in a completely different way than we saw in the case of $L_{\lambda C}^{\{S,T\}}$: what we have reconstructed as an expression of type A of $L_{\lambda C}^{\{S,T\}}$ within extensional semantics will now be reconstructed as an expression of type A/**I** of this new language. Montague's successors later took a similar approach to intensional logic: subsequent monographs on Montague's logic (Gallin, 1975; Janssen, 1986) translate Montague's original system into the language of Church's type theory thus modified; it is usually called the *two-sorted type theory* (Zimmermann, 1989).

Let us realize the difference between the intensionality of Montague's logic and the intensionality of Tichý's logic: in the first case, the intensionality is a matter of the formal system itself, while in the second it is a matter of how the system is used for natural language analysis. Hence, intensionality can be understood in two ways: as a purely formal property of a logical system, or as a property of the way we analyze natural language with this system. The first case corresponds to how intensionality is usually defined in textbooks of formal intensional logic: a logical system is called intensional if the law of intersubstitutivity of synonyms, and hence the principle of compositionality, does not hold within it. In this sense, Montague's logic is intensional (see the impossibility of getting the meaning of $^\wedge E$ from the meaning of $E$), but Tichý's is not. In the second case, we speak of intensionality when the meanings of sentences are explicated as functions from possible worlds to truth values; and similarly, the meanings of other expressions are also explicated as functions of possible worlds. In this sense, Montague's semantics is not intensional (let's remember that he always considers *extension* to be the meaning) while Tichý's semantics is.

# 5 Hyperintensional models of meaning: structure incorporated

## 5.1 Propositional attitudes and intensional isomorphism

We concluded that the extensional approach to semantics is unsustainable after finding examples of sentences that coincide in extension and yet are not interchangeable in some compound sentences without changing the extension of the whole. For example, (9) has the same truth value (and therefore extension) as (40), but if we replace the first of them with the second in sentence (38), we change the truth value (and thus extension) of this sentence from *Fa* to *Tr.*

(9) *Eco is a writer*

(40) *Eco is either a writer or not a writer*

(38) *Eco is necessarily a writer*

However, a similar argument may lead us to conclude that even intensional semantics, as outlined in the previous chapter, is not sufficient. Consider, for example, the sentence

(44) *John believes that Eco is a writer.*

Within sentence (44), can we replace sentence (9), which is part of it, with any other sentence with the same intension without changing the intension of the whole sentence? It seems, from the following example, that we cannot. Consider the sentences:

(45) $\pi$ *is irrational*

(46) *Eco is a writer and $\pi$ is irrational*

(47) *John believes Eco is a writer and $\pi$ is irrational.*

Sentence (45) is true; and because it is a mathematically provable sentence, it is true necessarily, i.e. in every possible world. (Sentences, the truth value of which is not a matter of empirical facts – especially mathematical sentences – are usually considered as keeping their truth values across possible worlds.[46]) The intension of this sentence (as well as the intension of any other mathematical truth) is therefore the set of all possible worlds. Sentence (46) is a conjunction of sentence (9) and sentence (45); it is therefore true precisely in those possible worlds in which both sentence (9) and sentence (45) are true. However, sentence (45) is, as we have just seen, true in all possible worlds, and hence sentence (46) is true in those possible worlds in which sentence (9) is true; in other words, sentences (9) and (46) have the same intension. Sentence (47) thus arises from sentence (44) by replacing sentence (9) with a sentence with the same intension (namely sentence (46)), and yet it seems clear that (44) can be true and (47) at the same time false: John can certainly believe that Eco is a writer without (if he is not too at home in mathematics) believing that $\pi$ is an irrational number. Sentences (44) and (47) thus may seem to have different truth values, and therefore different intensions, even if their components have the same intensions.

We will express it in a more concise way, which can be more illustrative. Sentences such as (44) and (47) express the relationship between a person and the meaning of a sentence, which is understood, within intensional semantics, as a set of possible worlds. However, because (9) and (46) express, as we have seen, the *same* set of possible worlds, sentences (44) and (47) say the same. (They both say that John is in a relationship of belief to the set of all the possible worlds in which Eco is a writer.) So how could one of them be true and the other false?

Sentences (44) and (47) are examples of so-called *belief sentences*; more generally the so-called *propositional attitude reports* (*X believes that Y*, *X doubts that Y*, etc.).[47] Thus, for such sentences, the principle of interchangeability of synonyms fails within intensional semantics; and this suggests that even intensional semantics, once we consider this type of sentences, is not fully satisfactory as a theory of meaning. Reflections on sentences of this type have prompted the search for semantic theories

---

[46] Though theories of possible worlds in which this does not hold were also proposed – see, e.g. Hintikka (1975).

[47] See Lewis (1972), Thomason (1980), or Peregrin (2000a).

that would be even "more intensional" than the standard intensional semantics; therefore, the theories arising from this search are called *hyperintensional*. It is not really a single semantic model, but a whole set of models that are sometimes very dissimilar (they differ much more from each other than, for example, Montague's and Tichý's intensional semantics, which we can still consider variants of the same model despite all the differences).

However, all the theories we call hyperintensional have one thing in common, and this allows us to summarize these theories under a common heading despite their considerable diversity. What they have in common is the belief that the meaning of an expression must be understood as a *structure* that is in one way or another due to the surface or syntactic structure of the expression. Probably the first explicit proposal in this direction came about in the middle of the previous century from Carnap; he suggested that genuine synonymy of two terms should not be explained as sharing the same intension, but as sharing the intensions of all the corresponding component words. Hence, two expressions are synonymous iff they consisted of the same number of words and the corresponding words have the same intensions. According to such a criterion, statement (9) would certainly not be synonymous with statement (46) (these statements consist of a different number of words), and this would explain why (44) can be true while (47) is false without violating the principle of compositionality. (According to this new criterion, sentence (9) could be synonymous at most, for example, with the sentence *Eco is a novelist* – but only if we recognize the words *writer* and *novelist* as synonymous.) Carnap (1947, §14) calls such a relationship *intensional isomorphism*.

Let us realize the essential difference between this understanding of meaning and intensional understanding. According to the intensional approach, the criterion of the identity of meanings is truth in the same possible worlds; and expressions that have a completely different structure can have the same meaning; for example, statements (9) and (46) (both have the intension of all the possible worlds in which Eco is a writer) or statements *One and one are two* and *Eco is a writer or Eco is not a writer* (these have for the intension the set of all possible worlds). According to hyperintensional approaches, the similarity of meaning involves a similarity of surface or syntactic structure – two sentences that differ in their structure as radically as (9) and (46) can therefore no longer be synonymous.

## 5.2 The "semantic structure" of an expression

We explained in the previous chapters the genesis of set-theoretical semantics, from the first suggestions of Frege to its fully fledged shape articulated by Montague and others. This development was driven by the idea that presenting meanings as set-theoretical objects is faithful to their nature, and besides this it is handy to work with. Considering the most elementary structures of natural language (or the structures of historically emerging languages of formal logic, which are used to model this language), we have come to the conclusion that we arrive at very clear and transparent models if we model the meanings of a substantial part of expressions of our language as functions (in the mathematical sense); the mechanism of composing the meanings of parts into the meaning of a whole thus became a simple functional application. This conclusion found its most general expression in our model of categorical grammar, which we then further enriched first with the mechanism of lambda abstraction and then, in the previous chapter, with intensions.

However, the problem of propositional attitudes calls into question the semantics built in this way: it seems that the meaning in the genuine sense of the word should not be any such object as a function, but rather something "structured". (If we follow Carnap's idea of intensional isomorphism directly, we will probably conclude that the meaning of an expression consisting of *n* words should be something like an ordered sequence of *n* intensions of these *n* words; more sophisticated elaborations of this idea lead, as we will see, to more complex structures.) A meaning seems to embody a structure that is somehow related to the syntactic structure of the expression by which that meaning is expressed, and this seems difficult to reconcile with the principles on which we have built our models so far.

This movement makes us return to the approaches to semantics that were here before the onset of set-theoretical semantics. (Of course, people studying natural language were engaged also in its semantic aspects in their own ways before that and kept being engaged in it after the onset.) Return to the "mathematical turn" of linguistics (see Section 2.5). When we talked about the advent of the Chomskyan approach to linguistics, we indicated that it proved appropriate to envisage the syntactic structure of an expression as a "tree"; such a capture of syntax also led to attempts to capture semantics by analogy. Couldn't such trees provide the structures we now seem to need for modeling meanings? To answer this question, we must first say something more about the structures that emerge in this context.

In Section 2.5, we gave a simple example of generative grammar and stated that a "tree" capturing the "derivation history" is considered the "syntactic structure" of this expression. Let us now give two more (simplified) examples:

(48) *Schwarzenegger admires Eco*

(49) *Eco is admired by Schwarzenegger*

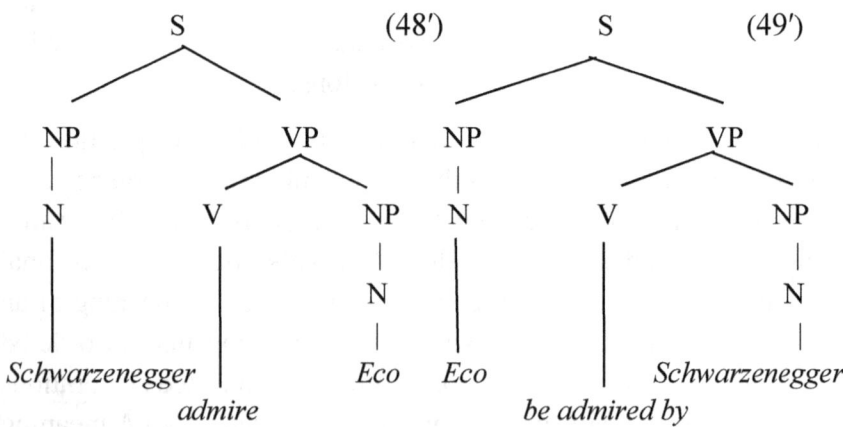

Within an alternative approach, based on the relationship of syntactic dependence (Sgall et al., 1986), the syntactic structures of sentences (48) and (49) could be captured – again in a simplified way – by trees (48″)

and (49″) (which, however, unlike those previous ones, do not capture anything like "derivation history").

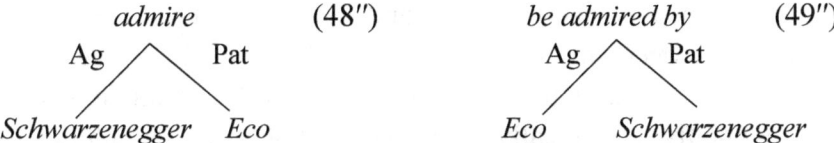

However, if we deal with sentences (48) and (49), we cannot fail to notice (even if we focus on syntax) that these sentences are in a sense equivalent, that they "say the same". This may lead us to the conclusion that in addition to the various "surface" syntactic structures of sentences (48) and (49), there is also a "deep" structure which is common to both sentences and which has to do not only with the syntax but also with the *semantics* of these sentences.

One possible response to this finding would be to say that the common structure is directly one of (48′) and (49′) or (48″) and (49″), and since the sentence in the active mood is usually felt as something more primary than the corresponding sentence in the passive mood, it will be (48′) or (48″). This would mean that in the case of (48) the deep structure coincides with the surface structure, while in the case of (49) it differs from it. Another possibility would be to declare, as a common deep structure of both (48) and (49), some completely new structure that would not be identical with either (48′) or (48″), or with (49′) or with (49″), but which would somehow capture what these structures "in depth" have in common.

Chomsky's theories and the theories of other linguists offer us a whole range of "deep" structures that are related to the meaning of the expression in one way or another. Chomsky originally worked on his concept of *deep structure* by dividing grammar into a generative part and a transformative part: according to him, the generative part generates a deep structure, which can then undergo various transformations before it becomes a surface structure. Thus, for example, a passive sentence (such as (49)) has the same deep structure as the corresponding active sentence (such as (48)), but it obtains a different surface structure by undergoing a transformation.

However, the considerations that led to the notion of deep structure have not yet been explicitly semantic (although implicitly often: we saw, for example, that the fact that the corresponding active and passive sentences share the same deep structure has to do with the intuition that the two sentences "say the same thing"). Lakoff (1971), for example, came up with an explicitly semantically motivated kind of "deep" structure within his theory of *generative semantics*; in Chomsky's conceptual framework, in which the number of different types of structures gradually emerged, it found its expression in what Chomsky calls *logical form* (which has little to do with logic).

However, from the viewpoint of a set-theoretic semanticist this was not really doing semantics but instead only beating about in the neighboring bushes. Really doing semantics consisted, according to theoreticians like Montague, in presenting the objects which are the meanings; and set theory offered a suitable framework for doing this. For these theoreticians, semantic interpretation was something that must be constituted under the auspices of the rigid rules of set theory (as we have seen in previous chapters); linguists unconnected to logic, on the other hand, used formal and schematic means quite freely, simply as a means of more clearly expressing what they would otherwise say informally about meaning. Chomsky and his followers captured the semantic structure using various tree schemes; for Montague's followers, this was not semantics in the true sense of the word. However, at the point we have now reached, a link between these approaches appears to surface.

Let us realize that the equality of "deep" structures (whether Chomsky's logical forms or other similar tools of linguistic theory) is a substantially more sophisticated concept than Carnap's intensional isomorphism. Carnap's criterion does not allow us to capture even a completely trivial synonymy of a simple expression with a compound expression (for example, the term *brother* with the expression *male sibling*), let alone cases of the synonymy of various syntactic constructions (*They fell silent after John's arrival* and *When John came, they fell silent*). Carnap and some of his followers sometimes expressed themselves as if they considered synonymy to be a problem that could be solved by a simple rule; on the other hand, it was clear to linguists that this was a problem that formed the core of semantics and must be the subject of an extensive theory.

## 5.3 Structured meanings

The synthesis of a rigorous logical approach to the formal capture of meaning with the Chomskyan approach was proposed in the early 1970s by David Lewis. Lewis (1972), one of the leading figures in Montaovian intensional semantics, makes it clear that capturing meanings of expressions using tree structures is not semantics in the true sense of the word; he sees it as only a translation of one language into another, namely natural language into a kind of a "tree language" ("Semantic Markerese"). However, the problems of intensional semantics and the need to understand synonymy as something closer than the sameness of intensions lead him to believe that Chomsky's theories could still be useful in this regard. (Lewis, however, argues with Katz & Postal (1964), whose approach is not directly related to Chomsky; but it is an approach that is close to Chomsky in the respect relevant here.)

The result of Lewis' reasoning is the connection between Carnap's idea of intensional isomorphism and a Chomskyan understanding of the meaning of an expression as a tree structure: Lewis proposed to capture the meaning as a tree whose end nodes are intensions. Thus, Lewis' proposal definitively abandons what could be called a *functional paradigm*: the assumption that the paradigm of how the meaning of a compound expression is given by the meanings of its parts is the functional application of one of those parts to the other.[48]

Thus, in Lewis' approach, if we continue to denote the intensions of the expression $E$ as $\|E\|$, the meaning of (9) will be something like (9'),

---

[48] The importance of the intensional model lies particularly in the fact that the paradigm, which seemed unsustainable, was essentially saved with the help of the notion of a possible world.

while the meaning of (46) will be something like (46′). Sentences (9) and (46) therefore have completely different meanings, and the difference in their meanings also explains the difference of meanings of (44) and (47).

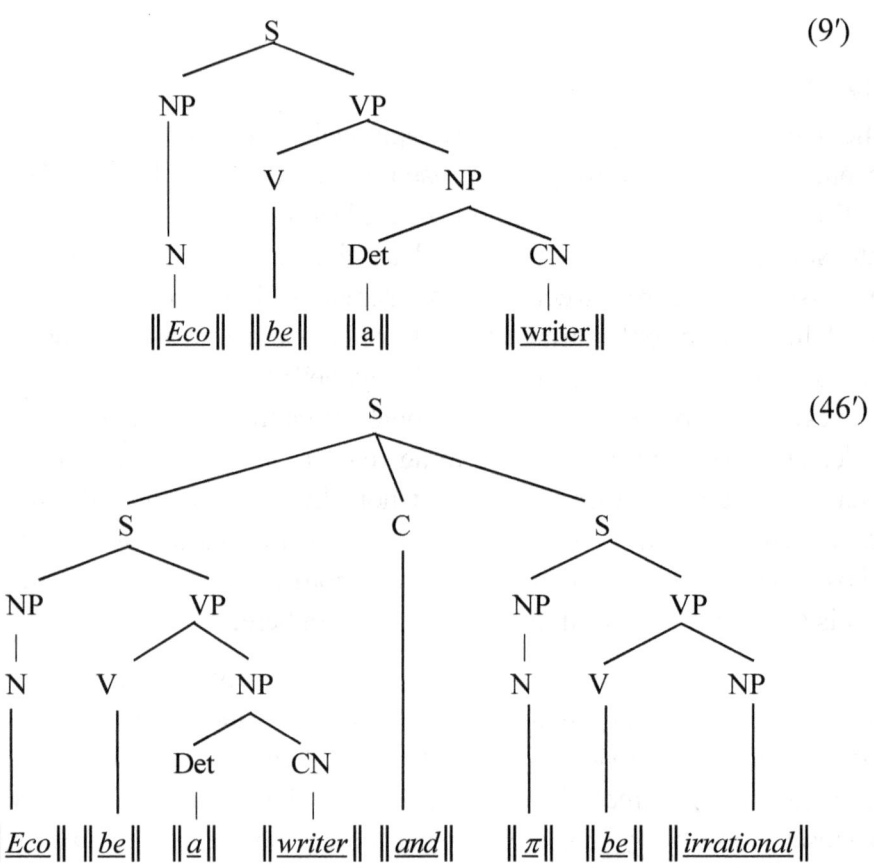

If the expression $E$ of category C resulted from combining the expressions $E_1, ..., E_n$ of categories $C_1, ..., C_n$, then if $.\cdot\cdot\cdot._i$ is the tree expressing the meaning of the expression $E_i$, the meaning of the expression $E$ is the tree

(46')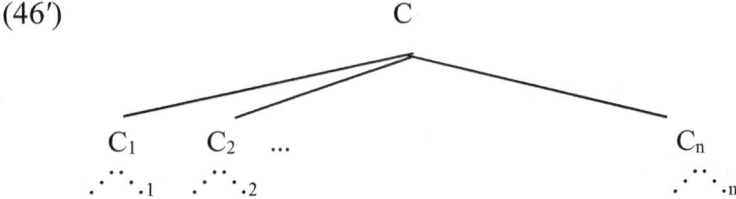

A special case is the meaning of the expression which arises from the categorical composition of the expression $E_1$ of category B/A with the expression $E_2$ of category A. It will be the tree

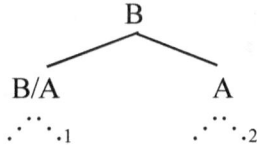

However, the nature of the theory of structured meanings and the success of how it deals with the problem of propositional attitudes depends, in a fundamental way, on what structures of expressions we take as its basis. If we based it on merely the surface structure, then it would be *de facto* simply a theory of the original intensional isomorphism of Carnap with all its shortcomings: two expressions would have different meanings whenever they consisted of different numbers of words, and different meanings would be assigned also to many expressions that otherwise serve as paradigmatic examples of synonymy (for example, the aforementioned sentences *They fell silent after John's arrival* and *When John came, they fell silent*).

For this reason, it seems quite obvious that the theory of structured meanings must be based on some deep structure, and Lewis's proposal thus essentially translates the problem into a problem of determining the deep structure. However, various types of theories of structured meanings have become part of the apparatus of many semanticists (see, e.g., Cresswell, 1985).

## 5.4 Are meanings still sets?

The set-theoretical paradigm that was embraced at the beginning of our extensional model was nice because many meanings reconstructed as sets did look plausible; for example, the identification of the extension of the word *dog* with the set of dogs looked quite natural, and so did that of their intensions as functions from possible world to classes of individuals. As a consequence, it didn't seem very problematic to think about meanings directly as about sets.

But is this not over now? Instead of seeing meanings as functions we see them as kinds of trees, complicated structures that might perhaps have to do with the implementation of meanings in the brain, but they do not look like meanings themselves. True, trees can be also formally understood as sets (for example, as ordered triples, whose components are a set of nodes which can be numbers, a set of pairs of these nodes, or edges, and functions that assign values to nodes and/or edges), But this is because any, or almost any, abstract entity can be grasped as a kind of set. Anyway, Lewis' proposal does not necessarily get you outside the framework based on the understanding of meaning as a set. However, it gets us certainly further from the original natural motivation of the set-theoretical notion of meaning.

Thus, the theory of structured meanings gives the impression of a somewhat inorganic combination of Chomskyan and Montagovian traditions. Pavel Tichý (1986; 1988) (whom we have already met, in Section 4.8, as the author of Transparent intensional logic) decided to face the situation head on. He was convinced that denotations of parts of a complex expression be combinable into the denotation of the whole by way of something as transparent as the functional application; yet he accepted the arguments leading to the conclusion that meanings are structured. His general conclusion, then, was that set theory cannot offer us objects of the kind we need and we must go "behind" it.

Tichý continues to rely on the concept of intension, abandoning only the idea that intensions are directly meanings. The intension of a compound expression is constructed from the intensions of its components, and Tichý is now convinced that the meaning of this compound expression is not the intension that is the result of this construction but this construction itself. According to him, the real meaning of sentence (9) is therefore not the set of possible worlds in which Eco is a writer, but the construction by which we get this intension from the intensions of the term *writer* and that of the term *Eco*. Tichý's hyperintensional conception of meaning is thus an organic superstructure of intensional semantics.

This proposal, needless to say, makes sense only insofar as we can make sense of the concept of *construction* understood in this way – and this is not easy. On the one hand, it is important to realize that constructions, according to Tichý, are not simply some structured set-theoretical objects, like sequences of intensions. The point is that constructions *do* something, namely construct objects (typically intensions) – and this is not something that can be meaningfully said about set-theoretical constructs. On the other hand, constructions are nothing like mere "computations", like executions of algorithms that transform some inputs into certain outputs. The constructions *contain* their "inputs" as their components. This led Tichý to conclude that his constructions are hitherto unknown kinds of entities that must be brought to light by semantic theory.

## 5.5 Tichý's constructions

Let us look at Tichý's general theory of constructing abstract objects from other abstract objects. What kinds of such constructions, according to him, can we recognize? We have already indicated that one type is the application of a function to its arguments, constructing the corresponding function value. An *n*-ary function and *n* objects would enter such a construction. However, Tichý assumes that it is not the objects that enter it directly but the constructions that construct these objects; the resulting construction is then called a *composition*. Thus, $n+1$ constructions enter into the composition, one of which constructs an *n*-ary function and the others construct the objects to which this function is applied within this construction. For example, if we have a construction that constructs the addition function, a construction that constructs the number 1, and a construction that constructs the number 2, we can compose them in a construction that will consist in applying addition to one and two, and we will thus construct number 3.

If we then have a construction, we can think of a function that arises when we make a "hole" in that construction that can be filled with various objects (similar to what we did with expressions when we introduced lambda-abstraction). Thus, such a "hole making" could be another type of construction, and the resulting structure would then construct the appropriate function. However, Tichý introduces this type of construction a little differently: he assumes that among the objects we are working with we have "variables" (but beware: these variables are not, like those we have encountered so far, expressions, but objects!) that are nothing but (objectified) "holes" that can be inserted into constructions.[49] This means that a "holey" construction can be created simply by using a variable in it instead of some construction constructing an object; and that construction corresponding to lambda-abstraction then, in fact, consists only in the fact that we begin to look at this "holey" construction as a corresponding function (and this is called letting it construct this function). Tichý calls this construction the *closure*. For example, if we have a construction that was mentioned at the end of the previous paragraph, we can replace the input construction that constructs the number 1 in it with a variable and we get a construction that constructs a function that assigns, to every number, the number plus 2.

The *composition* and the *closure* correspond to some extent to the two grammatical rules of our lambda-categorical grammar, i.e. the general language $L_{\lambda C}^{CAT}$ from Section 3.13 (however, it should be borne in mind that for Tichý these rules operate on objects or constructions rather than expressions). That's not a coincidence; Tichý believes that lambda calculus looks the way it does precisely because it expresses the most general principles of constructing abstract objects. These two basic constructions are also the most important in Tichý's system; the other three types of constructions that Tichý considers basic are already more

---

[49] It's a bit like Paul McCartney in the film *The Yellow Submarine*, when he picks up a hole that is on the ground, puts it in his pocket, and says, "I have a hole in my pocket!"

or less trivial.

The construction of an *execution* simply consists in taking an entity and, if it is a construction, then executing it, that is, constructing what that construction constructs; the construction of a *double execution* consists in doing the same and then, if the entity we constructed again is a -construction, executing it also. The last type of construction is *trivialization*, which is actually just a formal transformation of an object into a construction (note that all the constructions we have introduced so far assume that what enters them are constructions; trivialization is the only way an object can enter a construction). According to Tichý, this construction consists in taking the object and leaving it as it is.

Let us now define Tichý's system more rigorously:(0) *Variable*. A variable is a trivial type of construction; it is a kind of "objectified hole" that can be filled with various objects. We can imagine this construction as an act of selecting one object from a given universe. Variables are most often denoted by the letters $x, y, z, ....$ A variable constructs an object only relative to the valuation of the variables (assignment of values to them); then it constructs the object that is assigned to it by this valuation.

(1) *Trivialization*. Trivialization of X is a trivial construction of X from X; we take X and do nothing with it. The result of the trivialization is X, that is, what this construction constructs is X. The trivialization of X is denoted as $^0X$; for example, $^01$, $^02$ and $^0+$ are trivial constructions that construct the objects 1, 2 and +, respectively.

(2) *Execution*. If X is a construction, then the execution of X simply means that we execute X, i.e. we construct what it constructs; in that case, the execution of X is the same as X itself. The result of the execution of X in this case is the result of X. If X is not a construction, the execution of X is *improper*, it gives no result. The execution of X is denoted as $^1X$; that is, for example, $^1(^01)$ is a construction that constructs the object 1, while $^11$ is an improper construction, it constructs nothing.

(3) *Double execution*. The double execution of X means the execution of the execution of X, i.e. the execution of X, and if the result of this execution is again a construction, then the execution of this result. If X is

not a construction, or if its execution does not yield a construction, then this construction, i.e. the double execution of X, is improper. The double execution of X is denoted as $^2$X. Thus, $^2(^0(^01))$ is a construction that constructs the object 1, while $^21$ and $^2(^01)$ are improper constructions.

(4) *Composition.* If $X_0, X_1, ..., X_n$ are constructions and if the result of $X_0$ is an n-ary function, the construction of the composition of $X_0, X_1, ...$ and $X_n$ is executed by executing $X_0$ and applying the function constructed thus to the objects constructed by $X_1, ..., X_n$; the result is the functional value of this application. If any of $X_0, X_1, ..., X_n$ is improper or if $X_0$ does not construct an n-ary function, the whole composition is improper. The composition of $X_0, X_1, ..., X_n$ is denoted as $[X_0 X_1 ... X_n]$; so, for example, $[^0+ {}^01 {}^02]$ constructs the number 3.

(5) *Closure.* If Y is a construction and if $x_1, ..., x_n$ are variables, the closure of Y over $x_1, ..., x_n$ is a construction that constructs a function mapping the objects $a_1, ..., a_n$ on what would result from the construction of Y if the variables $x_1, ..., x_n$ constructed the objects $a_1, ..., a_n$, respectively. The closure of Y over $x_1, ..., x_n$ is denoted, as might be expected, as $[\lambda x_1 ... x_n Y]$; so, for example, $[\lambda x [^0+ x {}^02]]$ constructs the function that assigns every number the number plus two.

Tichý's theory is based on the intuitive assumption that an expression such as 1+1 does not simply express the number 2, it is an expression of a certain mathematical construction. That is, a sentence such as 1+1 = 0+2 tells us not the trivial fact that 2 equals 2, but that the construction of the sum of one with one leads to the same result as the construction of the sum of zero with two; and that is not so trivial. The denotation of a sentence like *Eco is a writer* will therefore be, within Tichý's theory of constructions, the construction of composition of the denotation of the term *writer* (which will be a trivialization of the intension of *writer*) to the denotation of *Eco* (which will be a trivialization of the intension of *Eco* – or, if we reject, together with Tichý, that proper names have intensions, the trivialization of its extension). It will not be, as it was in intensional semantics, *the result* of this construction, it will be this construction itself.

The answer to the question of what exactly is a construction is not, in fact, entirely unproblematic. Unlike the situations discussed in the next section, for example, it is not just a complex set; at the same time, however, it can hardly be a real execution of a construction process (that would mean that my addition of zero to two would be a different construction than the addition of zero to two performed by someone else). Constructions in Tichý's conception must therefore be imagined as a kind of "types of processes". They behave in one sense as structures or "conglomerates" (objects that enter them are preserved in them as their components), while in another as functions (they construct values). Here, however, we get into conceptual problems – normally, the general framework for defining abstract entities is set theory and, while Tichý does not intend his theory to be its part, he wants it to be something more fundamental than any formal theory.[50]

Compared to Lewis' theory of structured meanings, Tichý's theory is certainly much more elegant, and it fits much better with our intuitive notions of what a meaning might be: the claim that by a complex expression we express a tree on which intensions hang seems somewhat cumbersome and adopted entirely on an *ad hoc* basis; on the other hand, the claim that we express the construction of an object from the objects corresponding to the components of this expression does not seem to violate our intuition. This does not mean, however, that no problematic aspects can be found in Tichý's theory: for example, it is not very intuitive to say that the expression 1+2 expresses the construction of composition into which the constructions of addition, one and two enter; it would seem more natural to say that this expression expresses the construction of the addition into which one and two enter.

## 5.6 Situation semantics

The logician Jon Barwise and his colleague philosopher John Perry took a different approach to solving the problems that arise in semantics in connection with propositional attitudes when they developed, in the early 1980s, a semantic theory which they began to call *situation semantics*.

---

[50] See Peregrin (2000a).

This theory completely bypasses the notion of the possible world and replaces it, as the name suggests, with the notion of *situation*. What we typically talk about, and what the sentences of our language express are, according to Barwise & Perry (1983), situations (and not, for example, sets of possible worlds).

Unlike Lewis' or Tichý's theory, this theory derives the hyperintensional structures, which it declares as meanings of expressions, not from the structure of linguistic expressions (Lewis) or from the structure of cognitive performances of the human subject (Tichý), but from the (alleged) structures of reality itself. In doing so, it seeks to rehabilitate the original intuition that language semantics has to do with the relationship between expressions and what actually occurs in the wrld. This leads Barwise and Perry to believe that before we can deal with semantics, we must make a thorough analysis and inventory of the non-linguistic entities that make up our world and which can then be expressed in our language; hence. that before we develop situation semantics we must develop a general theory of situations. So they believe that before we can embark on semantics, we must deal with "ontology", and before we can create some formal *model* of semantics (in the sense we're discussing here), we must create some formal model of reality, some "formal ontology".

When we talked about possible worlds in the previous chapter, we mentioned that we can understand them either purely as a technical tool of semantic analysis or ("metaphysically") as something independent and scrutable independently of language. In the first case, the notion of a possible world is just a kind of expedient of the semantic analysis of modalities. The situation semantics of Barwise and Perry take the second of these paths (the notion of a possible world is, however, replaced by the notion of a situation); according to them, it is necessary to create a theory of how reality (from a human point of view) consists of situations, and semantics will then consist only in the appropriate pairing of expressions with elements of such reconstructed reality.

Situation semantics assumes that the key elements of reality, as perceived by language speakers, may be called *situations*. However, the basic building blocks of which situations consist are relations (of which properties are a special case) and objects. It is assumed that each relation has a number of arguments or roles; for example, a *reading* can be considered to have two roles corresponding to the person reading and what is being read (but we could also consider other roles, such as the addressee or the means of reading, such as glasses or a magnifying glass, the place of reading, or its time). If we assign a suitable object or person to each relation role, we get a specific instance (case) of that relation; such an instance of a *reading* relation is, for example, the assignment of the person of Eco and the book *The Hobbit* to their roles. Such an instance of a relation with added roles later became known as an *infon*. The infon can be supplemented by the so-called *polarity*, which can be plus (+) or minus (-): the polarity plus means that the objects contained in the infon really, "in the world", do play the roles in which they are occupied within this infon, while polarity minus means that this is not the case.

So, if we supplement *reading* with its roles occupied by Eco and *The Hobbit* with the plus polarity, we get an infon, which corresponds to the fact that Eco reads *The Hobbit*. The relevant infon is usually expressed by (50). The infon (50′) then indicates that Eco doesn't read *The Hobbit*.

(50) << *reading*, *Eco*, *The Hobbit*; + >>

(50′) << *reading*, *Eco*, *The Hobbit*; – >>

A set of infons is then what Barwise and Perry call *situation type*; and the situation type becomes a *situation* if we supplement it with spatio-temporal coordinates. (It is, more precisely, one of two kinds of situations, called the *state of affairs*; in addition to it, there are so-called *events* that do not relate to a single place and time but represent some spatiotemporal development.) So if we take a set made up by the single infon (50), it will be the type of situation in which Eco reads *The Hobbit*; if we add a specific time and place, we get one particular instance of a situation of this type, one particular situation in which Eco reads *The Hobbit*. However, we can also have a more complex situation, consisting of more than one infon, such as a situation in which Eco reads *The Hobbit*, drinks coffee (while reading), Schwarzenegger also reads *The Hobbit* (over his shoulder) and so on. Thus, one and the same infon can be contained in a number of situations – situations can differ in scope and can be nested.

The development of situation semantics following Barwise's and Perry's book then led to the introduction of infons in which some roles are unoccupied; more precisely, they are occupied only by formal objects called *parameters*. So we can have the infon (51), in which the first role is occupied by a specific object – Eco – and the second role by the parameter *x*; and we can also have the corresponding infon (51′) with opposite polarity.

(51) << *reading*, *Eco*, *x*; + >>

(51′) << *reading*, *Eco*, *x*; – >>

Infon (51) can be understood as representing the situation that Eco is reading something; while infon (51′) as that Eco doesn't read anything. Parameters, as we may think, play a role in situation theory similar to that played by standard variables in standard logic (but because they are not expressions but objects, they are closer to Tichý's objectualized variables than to variables of traditional languages of logic). Objects can be substituted for them; within situational semantics, assigning objects to parameters is called *anchoring*. There is also a mechanism corresponding to lambda abstraction – in situation theory it is called *absorption*. For example, by absorbing the parameter *x* in the infon (51) we get an infon (52) corresponding to the property *to be read by Eco*.

(52) [*x* | << *read*, *Eco*, *x*; + >>]

The fundamental difference, however, is that while the lambda-abstraction apparatus, as introduced in Section 3.12, was part of a (formal) language, within situation theory the apparatus of parameters and their absorption is part of what is expressed in that language. (Again, the variables and absorption of situation semantics are closer to Tichý's variables and the closure construction than to the variables and lambda-abstraction of categorical grammar.)

One can now expect that situation semantics will declare a situation or a situational type to be the denotation of a statement; however, Barwise and Perry make things more complicated by the fact that, in their explanation of meaning, they also try to deal with the problems of dependence of meaning on the context of speech. According to them, a sentence generally expresses a relationship between two situations: it relates the situation in which the sentence is pronounced to the situation which the sentence expresses. However, for some sentences, the situation of pronouncement will not play a nontrivial role: for example, the denotation of sentence (53) will be the relation that applies between *any* situation and the situation formed by infon (53), because what this sentence says does not directly depend on the situation in which it is pronounced; it's just true when Eco does read *The Hobbit*.

(53) Eco reads *The Hobbit*

In contrast, in sentence (54), the situation is more complex: its denotation will be a relation that will apply between two situations if a nd only if there is some *x* such that the first of these situations contains the infon << *speaker*, *x*; + >> and the second is formed by infon << *reading*, *x*, *The Hobbit*; + >>. This statement apparently says that someone reads *The Hobbit* and that someone is also the one who pronounces this sentence.

(54) *I read The Hobbit*

It is important to realize that Barwise and Perry's situation semantics represents a significant shift in the understanding of semantic theory compared to the theories discussed so far. On the one hand, it provides the same "mathematically precise" semantic model of language as the theories we have discussed so far, but on the other hand, it differs from these theories in that it does not rely directly on the procedures of formal logic. However, unlike those explicitly logic-oriented models, it can be understood as accommodating some semantic intuitions that have found no support in the models presented so far: intuitions that semantics has to do with something like "mental representations". For this reason, its main ideas deserve to be discussed in some detail. However, Barwise and Perry's theory is not widely accepted – despite the great interest it aroused in the early 1980s – and we will no longer talk too much specifically about it but rather more abstractly about the ideas on which it is based.

## 5.7 Situations vs. possible worlds

So far, we have talked about situation semantics as formulated by Barwise and Perry; let us now consider the concept of the situation on a more general level. We said that one of the basic motives for introducing situation semantics was an effort to rehabilitate the intuition that expressions express not some (quasi) linguistic or (quasi) cognitive structures, but the situations of our real world. This, of course, cannot be taken literally; we can certainly talk about what is not really, and if our sentences are to express situations, then they would have to be not only current (i.e. actually existing) situations, but also potential situations (which could exist, even if they do not actually exist). This means that we could generally see the meaning of a sentence as a possible, not necessarily real, situation (if we disregard Barwise and Perry's attempt to deal directly with the problems of the dependence of some utterances on the context, which makes them see meaning as a relation between situations). But the notion of *a possible situation* is not unlike the notion of *a possible world*: the difference seems to be only in the fact that situations are "smaller"; otherwise, situation semantics surprisingly approximates intensional semantics in this regard. So is there a significant difference between these theories? And is situation semantics hyperintensional at all, i.e. more "fine-grained" than intensional semantics?

We see one difference between these approaches when we ask ourselves what one understands when one understands the meaning of a sentence. In terms of intensional semantics, the straightforward answer would be to understand some set of possible worlds. This seems somewhat unnatural: if I understand a sentence like (54), I am confronted with something much more limited than a huge thing like a set of possible worlds. Even a single world seems to be something so vast that one could hardly contain it with one's mind; and the idea that such worlds are somehow crammed into their consciousness seems absurd at all. It seems far more natural to say that one understands the situation that corresponds to this sentence; it is this fact that largely justifies the appeal of situation semantics. (Note, however, that this argument is based on the idea that understanding a meaning is a literal "absorption" of the meaning into human consciousness or mind – and this is not a non-problematic idea.)

Situation semantics, in contrast to the semantics of possible worlds, thus seems appropriate to bring semantic theory closer to (cognitive) psychology and to accommodate the intuition that semantics has to do with "mental representations", with ideas about the outside world that are produced by our minds. Possible situations are (unlike possible worlds) easy to imagine just as mental representations of possible arrangements of things (which are then confronted with real situations). In this respect, situation semantics foreshadowed the later widely accepted (especially among linguists) *representational* approach to semantics, which we will discuss more in the following chapter.

Another difference between intensional semantics and situational semantics is that while intensional semantics usually doesn't tell us much about the nature of possible worlds, situation semantics tells us what situations are, what they consist of. However, this does not necessarily mean that situation semantics is more telling than intensional semantics. To say that the meaning of sentence (0) is the set of possible worlds in which Eco reads Hobbit is obviously trivial; but is it really less trivial to say that the meaning of this sentence is the situation that is formed by an instance of the relation of reading with its roles occupied by Eco and Hobbit?

Of course, it would be possible to add some non-trivial structure even to possible worlds, and some semanticists do it: Tichý, for example, says that a possible world is given by the distribution of some basic properties and relations among the individuals of the universe (which is thus the same for every possible world). To say that the meaning of sentence (53) is the set of possible worlds in which Eco reads *The Hobbit* is obviously trivial; but is it really less trivial to say that the meaning of this sentence is the situation that is formed by an instance of the relation of reading with its roles occupied by Eco and *The Hobbit*? With the right structuring, the possible world could then become a kind of maximal situation (which in itself contains smaller situations). In such a case, we could see intensional and situation semantics not as incompatible but differing only in that possible world semantics is the more coarse-grained in that it treats only "big" situations.

Aside from this kind of explanation of possible words as kinds of situations (the "big" ones), we can try it the other way around, i.e. to explain situations in terms of possible words. If worlds consist of situations, then we can say that the only classes of possible words that are available as meanings of sentences are the classes of all and only those worlds that contain a situation (i.e. sets of the form $\{w \mid s \in w\}$, where s is a situation); and such a set can then be seen as an equivalent of the situation their members share.

Hence, when intensional semantics says that the meaning of a statement is given by a set of possible worlds (in which that statement is true), situation semantics adds that it is always a set of worlds sharing a certain situation, and that it is better to take this situation directly into account. Thus, we can not only explain possible words in terms of situations, but we can also explain situations in terms of possible worlds; the gap between these foundational stones of different approaches to semantics narrows.

So if situation semantics can thus be directly linked to intensional semantics, is it hyperintensional at all? Can it deal with the problem raised in Section 5.1? This depends on whether it assigns the same or different meanings to sentences such as (9) and (46), and this in turn depends on how it interprets the sentence $\pi$ *is irrational*.

(9) *Eco is a writer*

(46) *Eco is a writer and $\pi$ is irrational*

At first glance, this may seem clear: its meaning will be the relation of *being irrational* with its only role occupied by the number $\pi$. But if we accept such a situation, then it will be a somewhat unusual one: it will be a situation that cannot fail to exist. However, if we accept this, then we will definitely separate situation semantics from intensional semantics – all mathematical truths will no longer be synonymous with each other, they will denote different situations.

## 5.8 From extension to situation

Let's go back to the extensional model for a moment. Within it, we understood names as denoting individuals (elements of the universe) and unary predicates (such as intransitive verbs) as denoting sets of those individuals about which the predicate in question is true. Binary predicates (such as transitive verbs), which we added to our model later in the chapter on extensional semantics, would then denote sets of pairs of individuals and similarly for predicates of greater arities. Therefore, if sentence (53) is true, we can illustrate it as follows:

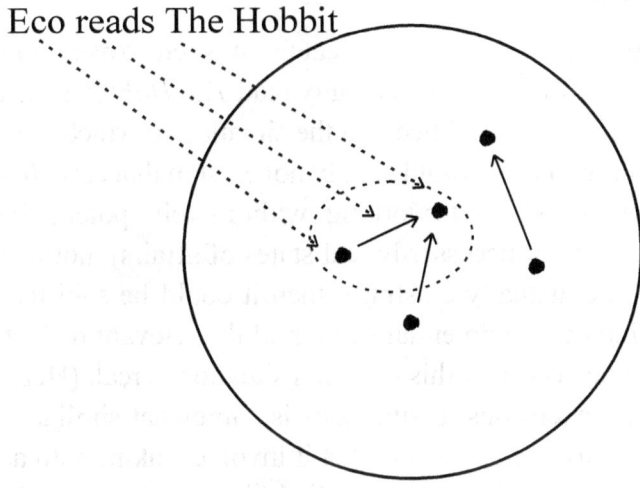

The names *Eco* and *The Hobbit* denote a certain two individuals, two elements of the universe, which we have shown in the picture with black circles. The binary predicate *read* denotes a set of pairs of individuals; in the picture we have shown this by means of arrows connecting the pairs of individuals that belong to it. (So an arrow from one individual to another indicates that the former reads the latter.) The dashed arrow from the predicate *reads* should then, strictly speaking, lead to all the arrows between the circles (the extension of *reads* is the set *of all* relevant pairs of individuals) but only the arrow connecting Eco with *The Hobbit* seems relevant for the interpretation of sentence (53), so we only lead the arrow from *reads* to it. Sentence (53) is true (in the context of extensional semantics it therefore denotes *Tr*) only when Eco is connected to *The Hobbit* by an arrow; otherwise, it is false (it denotes *Fa*).

There is thus a relatively small part of the universe which is relevant to the interpretation of sentence (53); we have encircled this part in the picture. It is this area of two individuals, Eco and the book *The Hobbit*, the former of which reads the latter, that would not be unnatural to see as the "situation" expressed in the sentence (53). Something like a situation is thus already implicitly contained in the extensional model; for extensional semantics to become a version of situational semantics it is enough to declare that it is these situations, and not truth values, that are really the meanings of statements.

However, there would be a catch in such "overcoming extensional semantics". If Eco did not actually read *The Hobbit*, sentence (53) would have no meaning at all because the situation in which we would consider the meaning of (53) would simply not exist in that case. It would therefore be necessary to start working with merely potential situations (i.e. possible, but not necessarily real states of affairs), not just real situations (current, i.e. actually existing); then it could be said that sentence (53) always indicates a potential version of the relevant real situation and that this sentence is true if this potential situation is real. (Here we see that the step from extensions to situations is somewhat similar to the step from extensions to intensions – in that it involves taking into account not only the mere "actual", i.e. what actually is, but also "potential", i.e. what could be.)

In this way, we would really get from extensional semantics directly to some simple variant of situation semantics. How would such a transition change our simplest extensional model, embodied in $L_E$ of Section 3.9?

The basic grammatical categories of this language are terms and predicates, which together form statements. We took terms as names of individuals from some set U (universe of discourse) and predicates as denoting something we will now neutrally call *relations* (within the extensional model they are sets of *n*-tuples of individuals). Let us denote the set of n-ary relations by $R^n$ (within the extensional model, therefore, $R^n = [Ux...xU \Rightarrow B]$). However, to get not only the *de facto* existing but all possible situations, we now need not only the real, *de facto* existing individuals, but also merely possible individuals, ones only potentially existing. Suppose, then, that U contains all possible individuals and that those that are real form only some subset of U. If then $r$ is an element of $R^n$ and $x_1,...,x_n$ are elements of U, we will call the (n+1)-tuple $<r,x_1,...,x_n>$ *possible situation* (PS). The set of real situations, RS, will then be a subset of the set PS of possible situations; that is, some of the possible situations will also be real (a possible situation will be real if the individuals it contains are real, and if those individuals also have the property that forms the first component of that situation).[51]

The semantics of $L_E$ by replacing points 3.1–3.3 and 4.1 with points 3.1' – 3.3' and 4.1'.

    3.1' The denotation of a term is an element of the set U.

    3.2' The denotation of a $n$-ary predicate is an element of the set $R^n$.

    3.5' The denotation of a statement is an element of the set PS.

    4.1' If $P$ is an $n$-ary predicate and $T_1,...,T_n$ are terms, then $\|P(T_1,...,T_n)\| = <\|P\|, \|T_1\|, ..., \|T_n\|>$.

As in defining denotations in this way we have lost the direct link between meaning and truth, we should reestablish it. The statement $P(T_1,...,T_n)$, whose meaning is the abstract situation $<\|P\|, \|T_1\|, ..., \|T_n\|>$ (an element of PS), is true just when the situation $<\|P\|, \|T_1\|, ..., \|T_n\|>$ is real (it is an element of RS).

Is it possible to proceed with the modification of the semantics of the language $L_E$? What about logical operators, quantifiers and statements that arise from their contribution? What is the situation denoted by a negative or disjunctive statement? Barwise and Perry introduced, as we have seen, "negative situations". This is somewhat unintuitive (to say that *Eco reads*

---

[51] Here we could again link such a theory of situations with the theory of possible worlds: we could say that by *a possible world* we mean any division of a set of abstract situations into real and unreal. This view also has something to do with the "metaphysics" of one of the most debated philosophical writings of the twentieth century, Wittgenstein (1922).

*The Hobbit* expresses that a certain situation is natural; but to say that *Eco does not read The Hobbit* also expresses a certain situation looks preposterous). Also, it only works for the negation of simple sentences. But what situation would the phrase *Eco reads neither The Hobbit nor The Three Musketeers* or its alternative expression *Eco does not read The Hobbit and does not read Three Musketeers* express – is it a situation different from *Eco does not read The Hobbit?* Or what situation would be expressed by the sentence *If Eco reads The Hobbit, then he can read*?

One solution, of course, would be to introduce disjunctive, implicative, etc. situations in addition to negative situations. However, in this way the notion of situation would lose much of its appeal. It gains this appeal from the fact that we imagine the situation as something that makes sense to us (at least to some extent) regardless of language, which is a kind of complex of things we see around us. (Such an extended notion of the situation might then also be largely similar to Tichý's notion of construction.)

If we insist on situations in this intuitive sense, what does a negative statement say? We seem to be able to answer that it expresses that the situation, expressed by the corresponding positive statement, does not really exist. The statement *Eco does not read The Hobbit* says that we will not find the situation expressed by the statement *Eco reads The Hobbit* in our real world. The disjunctive statement then says that there really is at least one of the two situations expressed by its respective disjuncts, and the implicative statement says that if there is a real situation expressed by its antecedent, there is also one expressed by its consequent. This means, however, that statements of this type do not directly express situations but rather express something *about situations*. This means that the equation *the meaning of a sentence = the situation expressed by the sentence* is generally not plausible (if we do not want to give the word situation a completely unnatural meaning). In general, it follows that semantics based on the idea that sentences express something like situations can look very natural and clever when we think of simple sentences (or their conjunctions), but it loses much of their appeal when we start thinking about negation, disjunction, etc.

## 5.9 Situation and mental representation

The modification of the language $L_E$ in the direction of situation semantics, which we proposed in the previous section, works with real situations as special cases of potential situations: some potential situations are real, others are not. Alternatively, we could imagine that potential and real situations are two different kinds of entities, with some potential situations *corresponding to* real situations.

If, then, we were to say *representations of situations* instead of *potential situations*, and we would say *represents a real situation* instead of *is a real situation*, we would have a formal framework for capturing what can be called *representational semantics*. This would change the image from the previous section as follows:

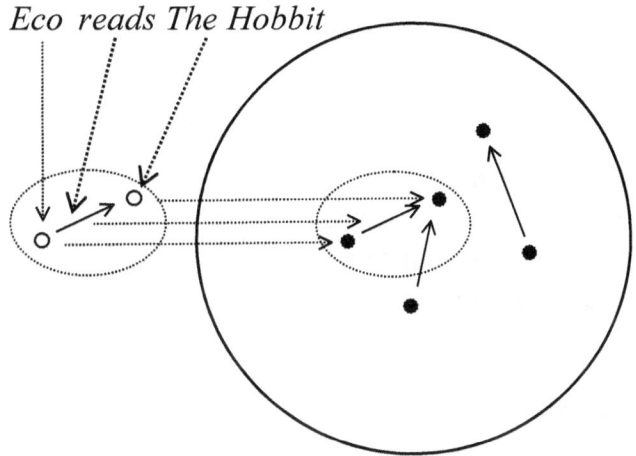

This would lead to a slightly different modification of the language $L_E$: we would leave the universe U as it is (i.e. it would consist only of real individuals), but in addition to it we would introduce a universe of *representing individuals* $U^R$ so that an element of $U^R$ may *represent* an element of U (but $U^R$ may also contain elements that do not represent any elements of U: for example, representations of individuals that do not really exist, or parameters). It could be similar with relations (in addition to the set $R^n$ of relations we could assume the set of potential relations or representations of relations) – however, unlike for individuals, this distinction does not make real sense for relations, so there is no need to duplicate them. If we take this approach, we can define that given elements $i_1,...,i_n$ of U and an element $r$ of $R^n$, then the $(n+1)$-tuple $<r,i_1,...,i_n>$ may constitute a *(real) situation*, and for elements $t_1,...,t_n$ of $U^R$ we will say that $<r,t_1,...,t_n>$ is a *representation* (of a situation). Furthermore, if $t_1,...,t_n$ represents $i_1,...,i_n$, respectively, we will say that the representation $<r,t_1,...,t_n>$ *represents* the situation $<r,i_1,...,i_n>$.

We would now modify the semantics of $L_E$ by replacing points 3.1-3.3 and 4.1 with points 3.1"-3.3" and 4.1":

3.1" The denotation of a term is an element of the set $U^R$ (of representations of individuals).

3.2" The denotation of an $n$-ary predicate is an element of the set $R^n$.

3.5" The denotation of a statement is a representation of a situation.

4.1" If $P$ is an $n$-ary predicate and $T_1,...,T_n$ are terms, then $\|P(T_1,...,T_n)\| = <\|P\|, \|T_1\|,...,\|T_n\|>$.

The link between meaning and truth will now look as follows: The statement $P(T_1,...,T_n)$, whose denotation is the representation $<\|P\|, \|T_1\|,...,\|T_n\|>$ of situation $<r, t_1,...,t_n>$, is true just when $<r, t_1,...,t_n>$ represents a real situation.

## 5.10 Where are meanings?

Could we present a formal language that would embody the principles of situation semantics in the form of our languages we presented earlier in the book? On the one hand, this would be no easy task: situation semantics is built on foundations very different from Montagovian intensional logic or other languages already discussed in this book, so to present it in this form one inevitably means some kind of violation. (An alternative possibility would be to present it in its original form in which, however, it may be quite confusing for readers of this book.) On the other hand, we have already put forward, in the previous section, two ways to articulate the core of such a language.

The most characteristic difference between these models and the previous languages appears when we characterize the denotations of statements. As the denotation of the predicate is nothing like a function, it cannot be *applied* to the denotation of terms; it is instead simply amalgamated with them: $\|P(T_1,...,T_n)\| = \{<\|P\|,\|T_1\|,...,\|T_n\|>\}$. (Let us keep in mind that our model is as simplified as possible; we work just with ordered *n*-tuples, whereas real situation semantics may help itself to a much more complicated structure.)

Could we complete our torso(s) of language(s) by the categories O1, O2 and Q plus the corresponding rules to accommodate logical constants and quantifiers? If we want to respect what Barwise and Perry say about them, it is not very easy. Let us start with negation.

Negation is directly connected with a constituent of situations, namely their polarity. We might want to say that negation sets the polarity to minus; but if already the situation expressed by the negated sentence may be minus, it would probably be more adequate to say that negation switches the polarity from plus to minus or minus to plus. All of this, however, concerns only sentences which express a situation consisting of a single infon. How to negate sentences which express situations consisting of multiple infons is not quite clear (though we can imagine how this could be done).

The functioning of conjunction looks rather transparent: a conjunction of two sentences expresses the union of the situations expressed by the two conjuncts. With respect to disjunction the authors claim that it is true if one of the disjuncts is. This looks like an ordinary logical disjunction. And implication doesn't occur in *Situations and attitudes* at all.

It may seem to go without saying that in whichever of the two above ways we understand situation semantics, the meaning of a statement will be the situation (or the representation of situation) expressed by the statement. But is this really so?

If we were to restrict ourselves to simple statements this might look unproblematic. A statement would express a situation, and it would be true if, and only if, this were a real situation or a representation of a real situation. However, statements are not only simple; there are also logically complex statements like negations, disjunctions or implications. Which situations would constitute their meanings? We have seen that it is, as a matter of principle, possible to build some logical complexity into the situations themselves. We saw that Barwise and Perry use the concept of polarity to distinguish between positive and negative situations, which would enable them to treat negation as related to something inside the situation. It would be possible to do the same with other operators to introduce disjunctive or implicative situations. But this would be on the verge of trivialization.

## 5.11 Autonomy of semantic structure

If we now think more deeply about the problems of semantic hyperintensionality and abstract from the individual theories, we can conclude that how we specifically understand "hyperintensions" (whether as a Lewis tree, a Tichý construction or a situation of Barwise and Perry) is perhaps less important than the kind of semantic structure we thereby impute to linguistic expressions. The semantic theories we have presented in this chapter are certainly incompatible if we take them as an answer to the question of what kind of an object meaning is (i.e. to the question of the "substantial" nature of meaning), but they are not necessarily incompatible if we take them as explanations of the semantic structure of the language, resp. its expressions.

From this point of view, we consider the semantic structure to be essential – rather than the idiosyncratic ways in which this structure is materialized in each of these approaches. It even seems that trees or other diagrams that do not meet what model-theoretically oriented authors expect from semantics could be sufficient to express this structure. If Lewis argues that the semantic theories of linguists are, in essence, not semantics but merely translations of natural language into "Marquerese", it is possible to argue that theories of model-theoretically oriented semantics are again in fact nothing more than translations, in particular translations into the language of set theory. Set-theoretical semantics is strong in that, within the functional paradigm (see Section 5.3), it has led to simple and transparent models (in which all combinations of meanings are realized through the application of functions to their arguments); however, if we fail to maintain this functional paradigm, as we have seen in this chapter, its merits become more debatable.

We can imagine that any semantic theory "classifies" sentences (and more generally expressions of any grammatical category) using a certain number of meanings and thus divides them into a number of "boxes", each corresponding to one of the meanings. (Each "box" thus contains mutually synonymous expressions.) We have just stated that a semantic theory is not so much a matter of the means by which it performs such "boxing" (i.e., what entities it declares to be meanings), but how fine-grained the "boxing" is (what it considers to be the criteria of synonymy). Extensional semantics has only two meanings for sentences (***Tr*** and ***Fa***) and thus divides sentences into only two "boxes". Intensional semantics leads to a much finer "boxing"; the problem is that, as it turns out, even this "boxing" is not fine-grained enough. For example, as we have seen, all mathematical truths fall into one and the same "box". The hyperintensional semantics thus makes the "boxing" even more fine-grained.

Thus, the semantics obviously progresses from a gross "boxing" to ever finer ones. The question immediately arises as to why not go straight to the "ultimate boxing" – there would be just one sentence in each "box". However, if we take semantic theory as a matter of revealing a semantic structure, such a semantic theory would be quite trivial – if each sentence had its own special meaning, we could stop talking about meanings at all and instead talk directly about the relevant sentences. What makes semantics non-trivial in this respect is that two different expressions can have the same meaning, that we cannot "read" the meaning of an expression from its surface structure. And we intuitively feel that semantics, even if it is understood structurally, is not trivial (the notion of meaning plays an important role in our pre-theoretical language).

Thus, it seems that a problem that is more fundamental than the problem of the nature of the capture of the semantic structure is what we might call the problem of the autonomy of this structure. On the path from extensional semantics to hyperintensional semantics we have limited this autonomy: while meaning understood as extension is completely independent of the structure of the expression that expresses it, the structure of meaning as we see it in this chapter is significantly affected by the structure of the expression. By limiting the autonomy of the semantic structure, we can deal with various intricate contexts of language (such as the contexts of propositional attitudes); now, however, we are already facing the problem of the complete liquidation of this autonomy and the subsequent trivialization of semantics.

We have already indicated that while set-theoretically oriented semantics thus moves from extensions to surface structures, more traditionally oriented linguists, on the other hand, have moved away from surface structures and, on the contrary, observe elementary semantic equivalences move away from surface structures. While the former have started from the minimal "boxing" and are constantly refining it, the latter have gone from the "maxima" and where possible, they simplify it.

Thus, the problem of the autonomy of the semantic structure seems to be the problem of articulating this counter-movement of semantics into criteria as rigid as the criteria of the original motion – then the "centrifugal" and "centripetal" forces can come into balance. As we have seen, we have clear criteria for when expressions necessarily *differ in* meaning – using these criteria, we have come to the unequivocal conclusion based on an analysis of modal statements that the meaning is not its extension, and then similarly based on the analysis of propositional attitudes, that the meaning of a statement is not even an intension. However, we do not have comparably clear criteria for when meanings necessarily *coincide*.

# 6 Dynamic models of meaning: contexts and utterances

## 6.1 Problems of anaphoric reference

Let us now return for a moment to the conceptual framework of extensional semantics. Sentence (55) would obviously be analyzed in it as (55').

(55) *Eco is a writer and Schwarzenegger admires him*

(55') <u>*writer(Eco)*∧*admires(Schwarzenegger,Eco)*</u>

However, sentence (55) is a combination of sentences (9) and (56), and (55') should therefore be the conjunction of the corresponding statements.

(9) Eco is a writer

(56) Schwarzenegger admires him

However, while statement (9) is quite straightforwardly analyzed as (9'), (56) alone can be hardly analyzed as (56').

(9') <u>*writer(Eco)*</u>

(56') <u>*admire(Schwarzenegger,Eco)*</u>

We can, of course, say that (55) is only an "abbreviated form" of the statement *Eco is a writer and Schwarzenegger admires Eco* and that (55') comes from this "complete" form; however, in this way we sweep some important semantic problems under the carpet. One problem is that the analysis of (55) is not compositional, in the sense that it cannot be obtained by a combination of the analyses of (9) and (57). Another one is that we wholly bypass any semantic characterization of pronouns such as *him*. Moreover, if we have sentence (58) instead of (55), the situation is even more complicated – in this case the pronoun *him* cannot be removed simply by taking the noun phrase to which this pronoun seems to apply (*someone*), and writing it in its place; nor by substituting *him* for something like *someone who is a writer* or simply *a writer*. In such a way we would get the sentence (59), and it certainly says something other than (58).

(58) If someone is a writer, then Schwarzenegger admires him

(59) If someone is a writer, then Schwarzenegger admires a writer

Within the extension model, sentence (58) corresponds to statement (58'), while sentence (59) corresponds to statement (59'); and while (58') is true only when Schwarzenegger admires every writer, (59') is true even if he admires only one of the many existing writers.

(58') $\forall x(\underline{writer}(x) \rightarrow \underline{admires}(Schwarzenegger, x))$

(59') $\exists x(\underline{writer}(x) \rightarrow \exists x(\underline{admires}(Schwarzenegger, x) \wedge \underline{writer}(x))$

What would work, it seems, is to use the phrase *the writer* to obtain (60) – but here the determiner *the* does the anaphoric work, so substituting it for *him* does not mean to explain it.

(60) If someone is a writer, then Schwarzenegger admires the writer

Intuitively, the situation seems obvious: a pronoun like *him*, unlike a name like *Eco*, does not always refer to the same object; it may refer to one object in one context and to another in another. In the simplest case, the object referred to by it is one that is picked up by a name immediately preceding the pronoun: so, if the sentence with *him* follows the sentence *Eco is a writer*, this *him* will refer to Eco, while if it follows *Schwarzenegger is an actor*, it will refer to Schwarzenegger. In other words, the phrase *Eco is a writer* will create a context in which *he* will refer to Eco, while the phrase *Schwarzenegger is a writer* will create one in which *he* will refer to *Schwarzenegger*.

However, to incorporate this intuition into a semantic model we must somehow incorporate *contexts* into the model. A straightforward way to do this would be to take recourse to Frege's maneuver once more: to make denotations of statements (and consequently perhaps also of other expressions) into functions of contexts. And we can think of the concept of situation from the previous chapter: the context could be perhaps understood as a situation, namely the situation expressed by the sentences that produced that context. Thus, sentence (9) expresses, according to situation semantics, the situation formed by the individual Eco with the property of writing; and this is exactly what we can consider as the context created by using this sentence. If sentence (56) follows, it already has this context "available", and it finds in it the individual who is to be referred to by *him*.

The problem of the analysis of pronouns, such as *he*, is then only a special case of the general problem of anaphoric reference; that is, the determination of what is referred to by an expression that is somehow affected by what is referred to by expressions used previously. For example, which person is referred to by the term *that person* is typically given by who has been spoken of before; while what time the term *then* refers to will typically depend on what time was talked about before. (What the anaphoric term refers to, however, can be determined not only by the previous text, but also by a non-linguistic context – to which person does the term *he* or the term *that man* refer can be given not only by who has been talked about so far, but also by pointing at somebody.)

However, all this necessarily leads to a new type of semantic model and, to a certain extent, to a somewhat different view of semantics. The considerations that led to our previous semantic models have always concerned expressions as abstract entities, disregarding the circumstances of their use. However, we now see that if we want to analyze expressions such as pronouns, we must turn our attention to specific linguistic utterances and to the way they follow each other. So we must think about what is called *discourse*: concrete coherent sequences of linguistic utterances. We must see a sentence not simply as something that expresses a meaning, but as something that enters a certain context, perhaps "uses" some elements of the context and at the same time modifies the context, thus making a new context that can then be used by subsequent utterances. Thus, we no longer look at language as something static, we also take its dynamics into account.

## 6.2 Articles

We have seen that for the semantic analysis of pronouns we need a semantic model incorporating *contexts*. But pronouns are not the only linguistic items which tend to be in this way context-dependent. Other words which are closely tied to contexts are articles as we find them in English or German. And as English articles got into the purview of logicians relatively early in the history of formal semantics, it is good to look at the early theories to see that they are not satisfactory and that we need new ones as introduced in this chapter.

But before we start discussing this, it is good to make a cautionary remark. We must realize that there are plenty of languages that have nothing like such articles. This is important to keep in mind especially when we believe that semantic analysis is something that leads us to objects or constructions that are already neutral to individual natural languages. If this were the case, then articles would have to be resolved during semantic analysis and no remnants of them would be present in the results. The fact is, though, that when we look at scholars who practice formal semantic analysis, we often see that they concentrate on English and that they do not shy away from using the articles as pillars of their semantic analyses. But having pointed this out we will mostly disregard it.

The English definite article is typically a tool of anaphoric reference; that is, something which must be tied to the context. If I use the expression *the dog*, I do not mean the only dog in the world, but the only dog salient in the current context. The indefinite article plays a different role. It introduces items to which we can later anaphorically refer. Typically using the phrase *a dog*, we introduce a dog into the current contents so that the next utterance of *the dog* can be taken to refer to it. Articles, then, similarly as pronouns and some other kinds of expressions, can be seen as tools for handling contexts.

A *locus classicus* of semantic analysis on the way to formal semantics is taken to be Russell's analysis of sentences with definite descriptions (Russell, 1905 – we have already touched upon it in several sections of this book). Russell's analysis was sophisticated. He argued that the sentence with a definite description, like *The king of France is bald*, says three things:

- that there is a king of France;

- that this individual is bald;

- and there is only one king of France.

The logical form of the sentence is then

(31') $\exists x(N(x) \land P(x) \land \forall y(N(y) \rightarrow (y=x)))$.

It is not difficult to extract the set-theoretic meaning of the definite article out of this (though Russell did not do it). We can introduce, as its formal counterpart, the Greek letter ι (originally it was the letter turned upside down, but as this turned out to be a nightmare for typesetters, this slowly gave way to the simplified version). Hence, if P is a unary predicate then ι(P) is a term. Thus ι(KF) may be the counterpart of *the king of France* and $\iota(\lambda x(KF(x) \land B(x)))$ the counterpart of *the bald king of France*.[52]

The meaning of ι is now a function that in the extensional case takes a set of individuals (like in our case the set of kings of France) and if this set is a singleton, i.e. if it contains exactly one individual, then it returns this individual. If it is empty (like the set of kings of France in the current world) or contains more than one individual, then it returns nothing and the whole sentence has no semantic value, i.e. is nonsensical.

Unfortunately, his analysis does not really work. In a great majority of cases when we use the definite article we do not mean the existence of a single individual in the world. When we say *the dog* we certainly do not presuppose that there is no more than one dog in the world.

---

[52] Russel's treatment of the iota operator was slightly different. He took it to be a variable-binding expression (as Fregean quantifiers) so that the counterparts would look slightly differently: $\iota x KF(x)$ and $\iota x(KF(x) \land B(x))$. In addition, Russell did not consider these terms as self-standing expressions but only as something that makes sense in the context of a sentence.

Therefore, we cannot take over Russell's analysis for our purposes as it does not reflect how the definite article really works.

What is interesting and much less well known than the Russellian analysis of the definite article is Hilbert's (1923) epsilon calculus, which has been later employed as an analysis of the indefinite article (Slater, 1991; Egli & von Heusinger, 1995). Similarly as in the case of the iota operator, if P is a unary predicate then $\varepsilon(P)$ is a term. Thus, $\varepsilon(KF)$ may be the counterpart of *a king of France* and $\varepsilon(\lambda x(KF(x) \wedge B(x)))$ the counterpart of *a bald king of France*.

The semantics of this operator, nevertheless, is more complicated than that of the iota operator. Hilbert let it be governed by a single rule, namely

if $P(T)$, then $P(\varepsilon(P))$.

This essentially says that if any individual has the property P, then the individual denoted by the term $\varepsilon(P)$ has that property. (In logic, the term *scapegoat* is sometimes used for individuals of this kind – its meaning becomes clear when we imagine that P is some unenviable property, such as being beaten. $\varepsilon(P)$ is then, we might say, the unfortunate individual who is guaranteed to be beaten whenever anyone is.) Given this, we can imagine that the term $\varepsilon(P)$ operates like 'a P': so, for example, we read $\varepsilon(writer)$ as 'a writer', and Charles = father($\varepsilon(writer)$) as 'Charles is the father of a writer'.

What is interesting is that if we take $\varepsilon$ as a fundamental operator, we can define both the quantifiers in its terms in the following way:

$\exists x P(x) \equiv_{Def.} P(\varepsilon(P))$

$\forall x P(x) \equiv_{Def.} P(\varepsilon(\lambda x(\neg P(x))))$.

This makes some sense if we start to read $\varepsilon(P)$ not just like *a P* but rather as something like *the most diehard P*. The prescription now says that there is a P iff the most diehard P is a P and it says that everything is P iff even the most diehard non-P is a P. This indicates that reading $\varepsilon(P)$ as *a P* is rather arbitrary.

What makes using ε as a counterpart of the indefinite article even more problematic than using ι as the counterpart of a definite one is the fact that ε defined in this way has no coherent semantics. (There are ways of producing semantics for it – see, e.g. Meyer-Viol (1995) – but this is not semantic in our fully-fledged sense.) From the fact that syntactically ε functions just like ι it would seem to follow that the extension of ε should be likewise a function from such of individuals to individuals. But which function? It is easily seen that any specific one of the available functions will do.

It follows that these attempts at formalizing articles have not been very successful. The reason, as I see it, is simple: articles cannot be accommodated except within the framework of dynamic semantics.

## 6.3 Discourse representation theory

Hans Kamp, a Dutch logician and linguist, came up with a theory based on the ideas outlined in the previous section in the 1980s; this theory became known as *discourse representation theory* (DRT) and soon became very popular, especially among linguists (Kamp, 1981; Kamp & Reyle, 1993). Kamp argues for the need for a completely new approach to semantics on the example of sentence (61) (these and similar so-called donkey sentences then became a symbol of the dynamic approach to semantics, just as unicorn sentences became a symbol of the intensional approach):

(61) If Pedro owns a donkey, he beats it.

This sentence consists of sentences (62) and (63) joined by the conjunction *if-then*. Here, (62) is straightforwardly analyzable as (62'), and as the variable $x$ appears in (62'), it could be expected, according to Kamp, that (63) should lead to (63'); and the whole sentence (61) to (61').

(62) Pedro owns a donkey

(63) Pedro beats it

(62') $\exists x(\underline{donkey}(x) \wedge \underline{ownPedro,x}))$

(63') $\underline{beat(Pedro,x)}$

(61') $\exists x(\underline{donkey}(x) \wedge \underline{own}(Pedro,x)) \rightarrow \underline{beat}(Pedro,x)$

But this is not the case: (61') is not an acceptable statement at all, because the last $x$ in it is already beyond the reach of the quantifier $\exists$; and if we extend this range as in (61''), we get a statement that says something other than (61).

(61'') $\exists x((\underline{donkey}(x) \wedge \underline{own}(Pedro,x)) \rightarrow \underline{beat}(Pedro,x))$

Indeed, if we take into account the semantics of the operator → (it holds that A→B is the same as B∨¬A; as can be immediately seen from the definitions of the relevant operators given in Section 3.4), it is easy to see that in order for this statement to be true, it is enough, for example, for there to be anything that Pedro would beat (it would not have to be his own donkey at all). The most correct extensional analysis of (61) is then obviously (61''').

(61''') $\forall x((\underline{donkey}(x) \wedge \underline{own}(Pedro,x)) \rightarrow \underline{beat}(Pedro,x))$

However, this analysis cannot be composed in a straightforward manner from the analyses of parts of (61) (i.e. from (62') and (63')). This means that there is something like a "failure of the compositionality of the analysis": we cannot analyze sentence (61), which is a sentence connected by *if-then*, by analyzing its parts and then combining their analyses using the operator corresponding to *if-then*, i.e. →.

Moreover, it seems to Kamp that the quantifier $\forall$ in (61''') is not acceptable: he feels that (61) says something about existence, not something general. We will shed more light on this objection when we note sentences (64) and (65):

(64) Pedro beats one of his four donkeys

(65) Pedro does not beat three of his four donkeys.

These two sentences are equivalent in the sense that one of them is true just when the other is true, and in the obvious sense they "say the same".

However, while we can follow (64) with (66), in the case of (65) this is clearly not possible; in this sense, these two sentences do not say exactly the same thing.

(66) We are sorry for it

The difference seems intuitive precisely in that (64) speaks of the existence of a single individual (and thus creates a context acceptable to the pronoun *it*), whereas in (65) this is not the case.

What solution does Kamp offer? He proposes to capture the "content" of sentences, but also of more complex units of discourse, by using the so-called *discourse representation structures,* which in the simplest case are nothing but certain types of situations as we generally discussed them in the previous chapter. In such a structure, all objects or individuals mentioned in the given sentence or discourse are captured as are the relations attributed to these individuals. For example, the content of statement (9) would be represented by a structure that can be written as (9″), and the content of the discourse formed by sentences (9) and (56) by structure (9+56″).

(9″)

| u |
|---|
| u = *Eco*<br>*writer*(u) |

(9+56″)

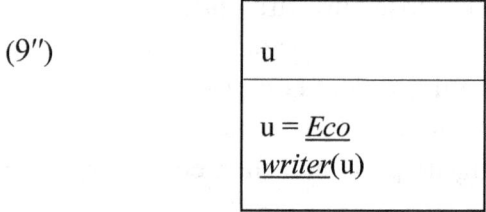

| u  v |
|---|
| u = *Eco*<br>*writer*(u)<br>v = *Schwarzenegger*<br>v *admires* u |

157

That this is nothing more than a situation is clear from how the discourse representation structure is formally defined. From a formal point of view, a discourse representation structure such as (9″) or (9+56″), is a set of individuals plus a set of instances of the relations between them. Thus (9″) is formed by a one-element set {u} and an instance *writer*(u) of the unary relation *writer*; and (9+56″) additionally contains an individual v and the instance u *admires* v of the binary relation *admire*. Thus, by means of situation semantics (see Section 5.5), we could capture the situation expressed as (9″) as (9‴) and that expressed as (9+56″) as (9+56‴).

(9‴) {<<writer, Eco; +>>}

(9+56‴) {<<writer, Eco ; +>>,

        <<admire, Schwarzenegger, Eco; +>>}

Kamp's approach can thus be seen as relying on representationally understood situation semantics. However, it differs fundamentally from this semantics in that it views its discourse representation structures dynamically: it is essential for it to consider how these structures change and evolve during discourse. When someone uses sentence (9), she "creates" a structure consisting of a single individual, who is Eco and who has the property of writing. If she continues to use (56), she "expands" this structure by another individual, Schwarzenegger, who is in the relation of admirer to the other individual. At the moment of interpretation of (56), we already have the structure (9′) and within it a set of "already mentioned" individuals to which pronouns may refer. And since structure (9′) contains a single individual, it isprecisely that which is a clear candidate for what *him* in (56) is to refer to.

A statement or discourse is then true if the structure it expresses corresponds to something in the real world. The discourse consisting of sentences (9) and (56) is therefore true if there are two individuals in the real world with whom u and v can be identified so that the relevant instances of the relations hold, that is, if there are two individuals, one of whom is Eco, who is a writer and is admired by the other, who is

Schwarzenegger. As in the case of representationally understood situational semantics, truthfulness is therefore understood within the DRT as the correspondence with reality.

At the same time, it is necessary to realize that, in general, a discourse representation structure can represent even more real-world situations. If we take, for example, sentence (62), then assuming that Pedro owns two donkeys, we will have two situations that we can see as represented by the respective representation (in both of them there is Pedro, but in each of them another of his donkeys).

However, everything is more complicated in the case of a statement such as (61). It is true if and only if every situation that is represented by the first of the pair of sentences that make it up (and we just saw that there can be more than one such situation) is part of the situation that is represented by both of these sentences. The relevant structure is captured by Kamp as follows:

(61″)
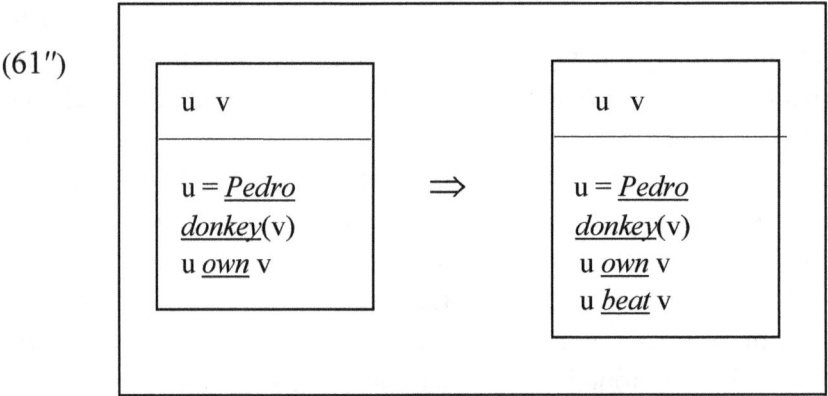

The sentence expressing this structure (i.e. (61)) will then be true if for each pair of real-world objects such that the first is Pedro and the second his donkey, it is the case that the former beats the latter. This captures what sentence (61) actually says.

Using such nested discourse representation structures, we can also analyze other sentences that require a general quantifier in standard logical analysis. Thus, the sentence (67) can be captured by the structure (67').

(67) Everyone is mortal

(67')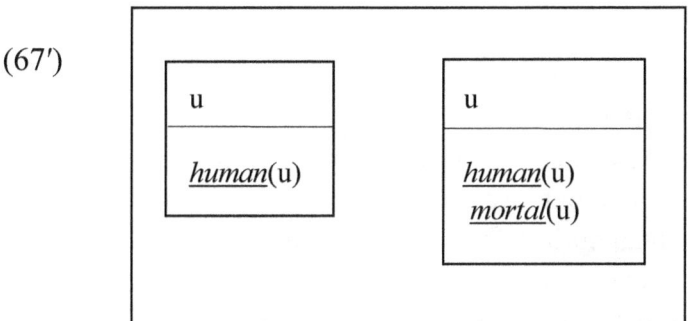

This already deviates from the straightforward notion that the discourse representation structure is simply a situation (in the ordinary sense of the word). (67') is clearly a more complex structure than a capture of a situation, it expresses that some situations are generally in a certain relationship. However, as we noted in the previous chapter when we discussed the limits of situational semantics (see the end of Section 5.7), if we want to semantically analyze not only simple sentences and their conjunctions, but also negations, disjunctions, implications, etc., we must begin to see what is expressed in sentences not directly as situations but rather as something *about* situations.

In sum: from the point of view of DRT, a sentence is something that contributes to building a representation of discourse, leading from one representation of discourse to another, "richer" (i.e. more specific) representation. Its denotation is therefore natural to understand as a process of rebuilding such a structure, or, what is more appropriate from a formal point of view, as a function that assigns a representation another representation, usually a "richer" one. Thus, the meaning of sentence (56) would be a function that assigns the representation (9+56") to the representation (9"). Here, therefore, the context into which the statement enters is understood as a representation of the recent discourse.

## 6.4 Meaning as a change of state

Let us now consider more generally the principles that form the basis of the "dynamic turn" of semantic theory that finds a possible expression in Kamp's DRT.

Each sentence is uttered in a certain context. The context is also sometimes considered as an *information state*, as a stack of knowledge that is considered as accepted by all parties of the discourse in question. (When we call it a context, we stress its function of a reservoir of potential items for anaphoric reference; when we call it an information state, we stress its role as a stack of shared knowledge.) An utterance of a sentence in an information state may change this information state in some way.

We can get a very general and very simple picture of how such a change of information state takes place using the framework of intensional semantics. We can imagine the information state as a set of those possible worlds that are still acceptable at a given stage of the discourse; we therefore understand discourse as the gradual exclusion of possible worlds. If I utter (9), I exclude all possible worlds in which Eco is not a writer; if I then utter (56), I further exclude all those in which Eco is not admired by Schwarzenegger. Such a view could lead us to take the denotation of a statement to be not a set of possible worlds, but rather a function that assigns to a given set of possible worlds another, usually more limited, set. By such a step, we would visually emphasize the dynamic nature of language, but we would not do anything revolutionary – our newly introduced denotation would be directly reducible to the initial one. If the standard, "static" denotation of the statement $S$ is the set $\|S\|$ of possible worlds, its "dynamized" denotation will be the function $f$ mapping every set of possible worlds on its intersection with the set $\|S\|$.

Semantics only becomes truly non-trivially dynamic not merely by the denotation of a statement becoming a function of the context or information state in which the statement is uttered, but when it is a function that is truly "sensitive" to some particular structure of that context; for example, when it gives a different result depending on whether that context contains an individual. This may be the case when we identify the context with something like a situation or Kamp's representation: as we have seen, for example, what statement (56) says is essentially dependent on whether the representation of the current discourse (i.e. the current context) contains an individual to whom the pronoun *him* could be related.

From the point of view of the problems of anaphoric reference, and these are what we concentrate on here, it is essential that the information state contains information about individuals who may become the subject of anaphoric reference. (So that whatever context or information state we model, it is likely to contain some "compartment" in which the recently mentioned, and therefore "salient", individuals will be placed. In DRT structures, these individuals are listed in the upper part.) This is what makes DRT dynamic non-trivially. The structure (9+56″) consists of two parts corresponding to (9) and (56): however, in order to build the part corresponding to (56), we must already have the one that corresponds to (9) and in it the individual Eco as the (only) candidate for the post to which the pronoun *him* in (56) refers.

In a way similar to DRT, the problem of anaphoric reference is dealt with by the so-called *file change semantics* proposed by Irene Heim (1983) at about the same time that Kamp proposed his theory. This theory works with a model based on the idea that by referring to an individual in discourse, a "file" is created for that individual, and thus that individual becomes a candidate for collaboration with subsequent anaphorically referential expressions; other references to this individual

can then add information to this file (and thus narrows down the scope of those anaphoric expressions that can refer to it), or close it and trash it.

However, it should be noted that neither DRT, nor file change semantics, solves all the problems associated with anaphoric reference in an exhaustive way (and we must keep in mind that the purpose of formal theory is not to solve everything in an exhaustive way – its strength is usually in that it neglects something to allow something else to stand out more clearly). In fact, the relationship between an anaphoric reference and context is extremely complex (what is anaphorically referenced to may have been mentioned before in a variety of indirect ways; different such ways lead to different probabilities with which the object may be referred to later, etc.). From this point of view, we would have to understand the structure of the information state in even more detail than via Kamp's representation of discourse or Heim's set of files. However, there seems to be a necessity of indirect proportionality – the less simplistic the theory, the less illustrative the model can provide.

## 6.5 Information states and their updates

As we see, the construction of a dynamic model of semantics is generally based on the concept of context or information state. We can see this concept as playing a similar fundamental role in dynamic semantics as the concept of the possible world within the intentional model. This means that just as we had assumed when building the intensional model, i.e. that we had a set of possible worlds (the nature of which we have moved out of our purview), we will now assume that we have some set of information states. However, a set of information states must be assumed to have (unlike a set of possible worlds) a certain structure: it is at least partially ordered, i.e. some information states are "richer" than other information states. If both $i_1$ and $i_2$ are information states, we will denote the fact that $i_1$ is part of $i_2$ (that is, in other words, that $i_2$ is an expansion of $i_1$) as $i_1<i_2$. Then we can further define terms such as the "sum" of information states: state $i$ is the sum of states $i_1$ and $i_2$ if: (i) $i_1<i$, (ii) $i_2<i$ and (iii) for every state $i'$ such that $i_1<i'$ and $i_2<i'$, it is the case that $i<i'$. The sum of the two states is therefore the smallest state that contains both.

Such a formal framework is then close to that developed to describe dynamic systems such as computers; the dynamized semantics of natural language thus, somewhat surprisingly, finds points of contact with the semantic theory of programming languages. The traditional understanding of the difference between natural language and programming languages can be expressed in such a way that while for natural language it is the "indicative mood" that is primary (the semantics of natural language is usually seen as based on indicative sentences), for programming languages it is the "imperative mood" (the language is seen as consisting of commands). This means that while natural language sentences indicate a current state, programming language commands aim at changing the state. We can then see the dynamic turn of natural language semantics as a move towards looking at natural language sentences more or less as instructions for changing an information state – namely the information state of the participants in the relevant discourse.

If we want to further analyze the concept of the information state, there are basically two ways, as is clear from what has been said before. One is to derive the notion of the information state from the notion of a possible world: the information state could be viewed as a set of possible worlds; and the statement $i_1$ *is part of* $i_2$ then read as *a possible world that does not belong to $i_2$, does not belong to $i_1$, so $i_1$ is a subset of $i_2$*. The second way is to identify the information state with something like a situation: this is, as we have seen, a way of DRT.

The basic characteristic feature of a dynamic model is that the statement indicates a function that assigns information states to information states. We will call such functions *updates*. The meaning $\|S\|$ of statement $S$ is therefore an update such that if $S$ is uttered in the information state $i_1$, it leads to the creation of the information state $\|S\|(i_1)$. Thus, if $\|S\|(i_1) = i_2$, we can say that by uttering (accepting) the statement $S$ we move from the information state $i_1$ to the information state $i_2$.

When we introduced the notion of a possible world, we justified this mainly by the need to define new, modal operators (such as □ and ◊) that were needed to analyze some natural language expressions and which could not be defined within the extensional model. The situation is similar in the case of the concept of the information state: in this case, the key operator is the one we can call *concatenation* and which we will denote with a symbol ⊕ (different types of symbols are used in the literature). The statement that is created by combining two statements by this operator will basically denote the composition of the respective updates, so it will be an update that assigns to the information state *i* the state that arises when we first apply to *i* the update corresponding to the first of the joined statements and then the update corresponding to the other:

$$\|S_1 \oplus S_2\|(i) = \|S_2\|(\|S_1\|(i)).$$

If the symbol * denotes the composition of functions (so *f*g* will be the composition of functions *f* and *g*, i.e. such a function that for every *x* for which *f(g(x))*, then we can see the operator $\oplus$ as a function that assigns their composition to two updates:

$$\|\oplus\| = \quad f, g \longrightarrow g*f.$$

It is then assumed that this operation corresponds to both the simple succession (i.e. pronunciation or writing in succession) of two natural language sentences and, in the typical case, their conjunction. If I state *Eco is a writer*, I will move from some initial information state to some other state, and if I then state *Schwarzenegger envies him*, I will move from this state again to another state. So if I state both of these sentences in a row, or if I state their conjunction *Eco is a writer and Schwarzenegger envies him*, I will go straight from the first to the last.

Note, however, that the operator $\oplus$ is not symmetric, i.e. $\|S_1 \oplus S_2\| = \|S_2 \oplus S_1\|$ may not always hold. If this operator is to be the "correct" analysis of the conjunction *and* instead of the standard, "static" conjunction $\wedge$ (as we have assumed so far), it would mean that even *and* does not behave symmetrically in natural language. But if we take into account statements containing anaphoric elements, then we see that this is indeed the case: saying *Eco is a writer and Schwarzenegger envies him* is certainly not the same as saying *Schwarzenegger envies him and Eco is a writer* (the latter obviously doesn't make good sense on its own). Intuitively, we can distinguish between statements that are somehow "context sensitive" and those that are not. For those which are context-sensitive, it is important which statement precedes them, i.e.

if $S_2$ is context sensitive, then apparently $\|S_1 \oplus S_2\| = \|S_2 \oplus S_1\|$ may not be the case.

What would then correspond to standard logical operators in terms of dynamic semantics? For example, if the statement $S$ denotes an update $\|S\|$, what would be the update $\|\neg S\|$ denoted by the negation $\neg S$ of the statement $S$? Informally, $\|S\|$ is an update from an initial information state to such an information state which, in contrast to the initial state, is richer in the information that $S$ holds. (However, if the initial state already contains this information, the update will not change it in any way.) The update $\|\neg S\|$ would then lead from the given initial state to a state that would be richer in the information that $S$ does *not* hold. To capture this within the dynamic model, we would probably have to give the information states a more non-trivial structure.

And if we limit ourselves to the analysis of sentence compositions, that is, if we remain at the level of propositional (albeit dynamic) logic and take statements as non-analyzed units, we cannot analyze the anaphora, which was the most important reason why we consider dynamic models of language semantics here. Anaphora is obviously a matter of parts of sentences – especially names and pronouns. To be able to say something about it on a very general level, we must therefore descend from the sentences to its parts; and a fruitful dynamic analysis of parts of sentences requires us, as it turns out, to say a little more about how we understand information states.

## 6.6 Information states as situations and as sets of possible worlds

As we have seen when we discussed DRT, one way of understanding contexts or information states is to identify these states with something like situations. This is based on the idea that discourse can be modeled semantically by modeling how the representation that is built by this discourse is constituted during the discourse. We have seen that an update of a discourse representation structure could be considered as the denotation of a statement within DRT, but what about parts of sentences like names and predicates?

As in the case of situation semantics, within DRT we can understand predicates as expressions of properties (or relations) and terms as expressions of (*abstract*) individuals. But now it is essential to distinguish between two types of terms: between those which introduce new individuals into the representation under construction and those that refer to already established individuals. The first will be those that match unspecific name phrases such as "someone", "a woman", etc.; the latter are those that represent pronouns and such noun phrases as "the woman". In addition to terms of these types, we can have, for example, terms corresponding to proper names that can name both individuals who are new in the discourse and those that have already been mentioned. It is the interaction between the terms of these species that is the essence of anaphora (if we leave aside the fact that in some cases the anaphora may also apply to entities other than objects). Thus, in the statement *A woman entered the room, and Schwarzenegger greeted the woman* the phrase *the woman* apparently anaphorically refers to the woman who was previously labeled (and thus "introduced") by the term *a woman*.

Another natural way to understand information states is, as we have also indicated, to understand them as sets, typically as sets of something like possible worlds. If we identify information states with sets of possible worlds, we identify them with what we understood in the context of intensional semantics as denotations of statements. Does that make any reasonable sense? An information state is a state in which some set of information is available; and because information is expressed by statements and the information communicated by a set of statements can be seen as well as communicated by a single long statement (that is, the conjunction of all elements of that set), the identification of the information state with the meaning of the statement is not incomprehensible. From the point of view of discourse analysis, we can see that the context in which a given statement comes is given by the conjunction of what has already been said in some discourse, and this conjunction defines the appropriate set of possible worlds. If we then utter the statement $S$, by which we enrich the available information, or by which we build on the existing discourse, we get a new (usually smaller) set of possible worlds. And within the dynamic semantics understood in this way, as we indicated at the beginning of Section 6.3, the meaning of $S$ will be the update from the first set to the second.

If we give the information states such a structure, we can classify the updates and thus the statements by which these updates are expressed in different ways. For example, the statement $S$ is said to be *static* if there is a set M of possible worlds such that for each set of possible worlds N, $\|S\|(N) = M \cap N$. (As follows from the considerations given in Section 6.3, if we dynamize intensional semantics only formally, the statements of the resulting system will be static in this sense; with $M = \|S\|$). The statement $S$ is called *distributive* if for every set of possible worlds M it holds that $\|S\|(M) = \bigcup_{w \in M} \|S\|(\{w\})$, that is, if the set of worlds that we get to by the update $\|S\|$ of M is the same as the union of the sets of worlds that we reach by the updates of the individual elements of N. This update is, in other words, distributive, if to know where it leads from some sets of possible worlds it is always enough to know where it leads from each individual world of this set. If the update is distributive, we can understand it simply as a relation between possible worlds: if we define the relationship $R_{\|S\|}$ by the rule

$w \, R_{\|S\|} \, w'$ just when $w' \in \|S\|(\{w\})$

(the world $w$ will be in this relation to the world $w'$ just when we can get from $w$ to $w'$ by the update $\|S\|$), then we can reconstruct $\|S\|$ from $R_{\|S\|}$ at any time, because for each M there will be

$\|S\|(M) = \{w \mid \exists w' \in M : w' \, R_{\|S\|} \, w\}$.

This means that so long as we limit ourselves to distributive updates, we can from a formal point of view – instead of updates between sets of possible worlds – simply talk about the relations between possible worlds. A distributive statement $S$ is called a *test* if $\|S\|(M) \subseteq M$ holds for each set of possible worlds. Thus, a statement is a test if it does nothing but exclude some of those possible worlds which are still possible at a given stage of discourse. Every static statement is a test, but a test generally does not have to be static.

If we restrict ourselves to distributive updates, there comes to the surface a surprising similarity between the semantics of modal logic and our dynamic semantics. In Section 4.2, we noted that the general semantic framework for modal logic, proposed by Kripke, consists of a set of possible worlds and an accessibility relation between the possible worlds; this accessibility relation is what establishes modalities such as possibility and necessity. From a formal point of view, therefore, this framework is simply a set and a certain binary relation on this set; whereby statements are assigned subsets of this set. However, we can also consider frameworks with several different binary relations (these are needed, for example, for the interpretation of so-called multimodal logics, for which we then have different kinds of "possibilities" or "necessities"). And since dynamic propositional logic can now be interpreted essentially within frameworks of the same kind (but with the significant difference that statements are now assigned not sets of possible worlds but binary relations between worlds), there is a formally interesting relationship between modal and dynamic logic (what is the meaning of a statement, from the point of view of dynamic logic, is, from the point of view of modal logic, the basis of modalities).

However, indices whose sets are information states do not necessarily have to be understood as real possible worlds. For example, if the only thing we are interested in is the analysis of the semantics of pronouns, we can, for simplicity, look away from the state of everything else and understand the relevant index only as the state of potential values of pronouns. The "possible world" is then created for us only by what value in a given state belongs to which pronoun: in one such "possible world" *it* will potentially be associated with Eco (such a state would arise, for example, after the sentence *Eco is a writer*), in another it is associated with Schwarzenegger (for example, after the statement *Schwarzenegger is an actor*) and in another with nothing at all (for example, if nothing has been said yet).

## 6.7 Dynamic predicate logic

Dutch logicians Jeroen Groenendijk and Martin Stokhof based their dynamic predicate logic (DPL), which was the first (however intentionally very primitive) truly logical model of anaphora, on this idea (Groenendijk & Stokhof, 1991). They introduced a new type of expression – the so-called *discourse markers* – which represent anaphorically functioning terms, i.e. something like formal pronouns. Although these expressions seem similar to variables of traditional logic, to consider them as variables would be completely misleading – their actual function is substantially different, having to do with anaphora. Within the DPL, the denotation of a statement can be understood as the update of sets of such assignments of values (universe elements) to discourse markers. (Perhaps we can see one such assignment as a "possible world"; and an information state as a set of such "possible worlds".) However, because the updates that DPL works with are distributive, we can understand its statements (based on what has been said above about distributive updates) as expressing the relations between the assignments of values to discouse markers.

DPL, as proposed by Groenendijk and Stokhof, is therefore an example of what is probably the simplest way to reconcile logic with the kind of dynamics needed to analyze anaphora. Its language is in fact syntactically identical to the language of standard logic (that is, to our simplest extension language $L_E$ from section 3.7); however, with the significant difference that what variables are for traditional logic are discourse markers for DPL; and as a result, the principle and mode of operation of some seemingly standard DPL operators are fundamentally different.

A DPL statement that contains discourse markers is not true or false by itself – it becomes true or false only when some values are assigned to the discourse markers (and such an assignment is given by the current context). If we call such an assignment a *valuation*, a DPL statement will generally be true only relative to a valuation (similar to a standard logic formula that contains free variables, true only relative to assigning values to those variables). The DPL statements will then indicate a relation between valuations (hence the updates from valuation sets to valuation sets) and the statement S will be true with respect to a given valuation $f$ if there is a valuation $g$ to which we can get from $f$ through S; i.e. if $<f,g> \in \|S\|$ for some $g$.

The operator $\wedge$ of the DPL language is not the classical conjunction, but rather a "concatenation"; that is, the sign $\wedge$ in the DPL denotes what we have identified in Section 6.4 as $\oplus$. Statement $S_1 \wedge S_2$ thus leads from the valuation $f$ to the valuation $g$, just when there is some valuation $h$ such that $S_1$ leads from $f$ to $h$ and $S_2$ leads from $h$ to $g$. The statement $S_1 \rightarrow S_2$ always leads from the valuation $f$ at most to the same valuation $f$ (i.e. it cannot change the valuation), just when for each valuation $h$ such that $S_1$ leads from $f$ to $h$, $S_2$ leads from $h$ to $g$.

The most fundamental non-standard element of the DPL language is the dynamic existential quantifier $\exists$; it is the existentially quantified statements that are the basic statements that change the context. $<f,g> \in \|\exists x S\|$ just when there exists a $h$ which differs from $f$ at most in the value assigned to $x$ and $<h,g> \in \|S\|$. That is, $<f,g> \in \|\exists x S\|$ only if $h(x)$ is an object that satisfies S in the sense of standard predicate logic – and for each statement that follows $\exists x S$, then $x$ is necessarily associated with some such object. Seen through the lens of traditional logic, an existential quantifier can bind variables (in dynamic logic, of course, these are not variables, but discourse markers!) even outside their normal scope. The genral quantifier, on the other hand, is defined essentially in accordance with standard logic.

In studying the following definition of the language of Groenendijk's and Stokhof's dynamic predicate logic, keep in mind that this is intentionally the simplest possible version of logic that would allow the anaphora to be analyzed; therefore, it cannot be taken as an adequate model of natural language semantics.

To characterize the language of dynamic predicate logic, we can say that it has six categories of expressions: **T** (an unlimited number of simple constant terms and an unlimited number of discourse markers), **P$^n$** (an unlimited number of simple n-ary predicates), **O1** ($\neg$), **O2** ($\wedge$, $\vee$,

→), **Q** (∃, ∀), **S** (no simple expressions). In addition, we assume that we have auxiliary, syncategorematic symbols: parentheses.

There are four syntactic rules: if $P$ is an n-ary predicate and $T_1,\ldots,T_n$ are terms, $P(T_1,\ldots,T_n)$ is a statement; if $O$ is a propositional operator and $S$ is a statement, $OS$ is a statement; if $O$ is a propositional operator and $S$ and $S'$ are statements, then $SOS'$ is a statement; and if $Q$ is a quantifier, $x$ is a discourse marker and $S$ is a statement, $Qx(S)$ is a statement.

Denotations of simple expressions are as follows: the denotation of a term is an element of a given set U; the denotation of an n-ary predicate is a function from U×…×U to B, i.e. an element of [U×…×U ⇨ B]; we do not talk about the denotation of unary propositional operators (although it would be possible to identify the denotation with the function from valuations to valuations); we do not talk about the denotation of binary propositional operators (although it would be possible to identify the denotation with the binary function from valuations to valuations); the denotation of a quantifier is the function from [U ⇨ B] to B, i.e. the element [[U ⇨ B] ⇨ B]; the denotation of the quantifier ∀ is the function that assigns the value **Tr** to the set M just when M = U; the denotation of ∃ is the function that assigns the value **Tr** to the set M just when M is nonempty.

And the prescriptions for computing denotations of complex expressions from those of their parts are as follows:
$\|P(T_1,\ldots,T_n)\|$ = {<g,g> | $\|P\|(\|T_1\|g,\ldots,\|T_n\|g)$ = **Tr** }, where $\|T\|g$ is $\|T\|$, if $T$ is a constant term and $\|T\|g$ is $g(T)$, if $T$ is a discourse marker.

$\|\neg S\|$ = {<g,g> | for no $h$ <g,h>∈$\|S\|$}

$\|S \wedge S'\|$ = {<g,h> | for some $k$ <g,k>∈$\|S\|$ and <k,h>∈$\|S'\|$}

$\|S \to S'\|$ = {<g,g> | for every $h$ such that <g,h>∈$\|S\|$, exists for such that <h,k>∈$\|S'\|$}.

$\|S \vee S'\|$ = {<g,g> | for some $h$ either <g,h>∈$\|S\|$ or <g,h>∈$\|S'\|$}.

$\|\exists x(S)\| = \{<g,h> \mid$ there is a valuation $k$ such that $k(x') = g(x')$ for each discourse marker $x'$ other than $x$, and $<k,h> \in \|S\|\}$

$\|\forall x(S)\| = \{<g,g> \mid$ for each valuation $k$ such that $k(x') = g(x')$ for each discourse mark $x'$ other than $x$, there exists a valuation $h$ such that $<k,h> \in \|S\|\}$

Within dynamic logic, (9) and (56) would be schematized as (9'''') and (56'''').

(9'''') $\exists x(\underline{writer}(x) \wedge (x = \underline{Eco}))$

(56'''') $\underline{admire}(\underline{Schwarzenegger}, x)$

The fact that (9'''') is not simply $\underline{writer}(\underline{Eco})$ is due to the fact that in DPL the anaphora is mediated by purely discourse markers, and the marker $x$, which represents the pronoun in (56''''), must also occur in (9'''') in place of the name that names what the pronoun in (56'''') is to refer to. The "effect" of this analysis can then be approximated so that the interpretation (9'''') "lets through" only those evaluations of discourse markers that assign the value Eco to $x$, and (56'''') is thus interpreted how it should be.

We have noted that a dynamic view of natural language semantics leads to a certain convergence of this semantics with the semantics of programming languages. In the case of DPL, this general parallel is even more pronounced. Because the state of a computer is determined by the program's values of its variables, the natural meaning of the command is a binary relation that takes an assignment of values to variables to s (possibly different) assignments of values to variables. By analogy, the DPL then takes a natural language sentence as something that changes the assignment of values to anaphorically referring expressions: that is, as a binary relation between such assignments.

# 7 Meaning as an object

## 7.1 Formal semantics

In the preceding chapters we discussed and analyzed, in detail, how meanings can be presented, and have been presented, as set-theoretic objects. A question we can ask is to what extent is it adequate to capture all meanings as objects. Though there are semanticists who would vigorously agree with this, there are certainly also those who would disagree. Another question is whether meanings can be adequately captured as *set-theoretic* objects. In the previous chapters we have tried to indicate what makes sets so particularly suitable to play this role. And we have also tried to elaborate the question closely related to the previous two: how do we construct – or discover – the kinds of sets which can embody the meanings of the various kinds of expressions of our languages?

And now *the* philosophical question: What does capturing meanings as objects tell us about meaning? Is it desired simply because meanings *are* objects? Or is it desirable for some other reason, perhaps having to do with a methodology of explanation? Or is it, the other way round, misguided?

The conviction that meaning is an object is often interconnected with what I called the semiotic view of language (Peregrin, 2001): the view of the meaning of a linguistic expression as an object represented by the expression. (This is a rather widespread view, which goes back at least to Peirce.) Consider the celebrated depiction of the relationship between language and reality as given to us by Wittgenstein (1922, §3.2, 3.22, 4.01):

> In propositions thoughts can be so expressed that to the objects of the thoughts correspond the elements of the propositional sign. ... In the proposition the name represents the object. ... The proposition is a picture of reality.

Or consider Reichenbach (1947, p. 4):

> Language consists of signs. ... What makes them signs is the intermediary position they occupy between an object and a sign user, i.e., a person. The person, in the presence of a sign, takes account of an object; the sign therefore appears as the substitute for the object with respect to the sign user.

Thus, at least some of the twentieth century pioneers in semantics presumed that linguistic expressions were signs, and what made them signs was that they stood for objects. This assumption is built into the foundations of many theories of meaning, even some of those which do not subscribe to it explicitly. Hence, is the model-theoretic view of semantics that treats meaning as set-theoretical objects to be read as a reinforcement of this semiotic view?

Returning to history, of course, not everybody subscribed to this view. Some semanticists, like Frege (1891; 1892a), insisted that only some expressions represent objects (the saturated ones, i.e. names). Other kinds of expressions (the unsaturated ones) denote functions (which, according to Frege, were not objects). And then there are the defenders of various "use theories" of meaning, like the later Wittgenstein, according to whom meanings are not something that are stood for by expressions, they are rather the ways the expressions are employed by their users.

Hence, there were always theoreticians of language who were convinced that meanings are not objects, and that expressions are not really signs. This view of language gained support during the later twentieth century, in connection not only with the later Wittgenstein, but also with the American philosophers like Quine (1960), Sellars (1974), or Rorty (1979). As Wittgenstein (1953, §11) puts it: "Think of the tools in a tool-box: there is a hammer, pliers, a saw, a screwdriver, a rule, a glue-pot, glue, nails and screws.—The functions of words are as diverse as the functions of these objects."

Thus, the philosophy of language in the later twentieth century was marked by the opposition between two different general approaches, which I call the *label approach* and the *toolbox approach*. According to the former, meanings are, first and foremost, *representations* ("labels"); according to the latter, they are especially the means of our *interaction* with each other and with the world ("tools"). In any case, all of this substantiates an inquiry into the general question: *is meaning an object?*

## 7.2  Being an object vs. being conceived as an object

We saw that versions of set-theoretic semantics have flourished; but before evaluating what they have brought to us, let us consider the alternative we mentioned earlier. What if we reject that meanings are objects at all? Perhaps our speaking about them as objects should be seen merely as a *façon de parlor*, like we sometimes talk about the lengths of things though we know that we need not take this talk as more than simply our treating of the things being variously long. In the same way, it would seem, we may sometimes talk about the meanings of expressions without necessarily taking this as more than our treating of the expressions being variously meaningful (which may further mean that they are various kinds of vehicles of our language games).

Note, however, that even if we opt for this view, it may still be possible to grasp meanings as objects. We have no problem saying that different sticks have different lengths, though we know that this is just a way of talking about how long they are. Likewise, speaking about meanings as about objects, though we do not intend it to be taken literally, also need not be problematic – especially when this turns out to be useful.

In this case, we feel the urge to say that lengths or meanings are not *really* objects, that they are merely *conceived* as such. And the question then becomes what would it take to *really* be an object, especially what would this take for meanings. Let us return to the discussion from the first chapter of the kinds of objects meanings can be. We rejected the possibility that meanings can be spatio-temporal objects because of the problem of scarcity, and we also rejected the possibility that they can be an object of a mental realm because of the problem of intersubjectivity. Hence, we cannot say that meanings are really objects if and only if they exist in space and time; we also cannot say that they are really objects if they can be thought about as objects. So how can one distinguish between cases where meaning is *really* an object and cases where it is merely being *taken as* an object? And is there really any clear distinction? It seems that to answer these questions we must widen our focus to capture not only meanings, but also the broader context of semantic theory.

It is, therefore, difficult to answer the question whether meaning is really an object – due to the unclear sense of the "really". But there is a different, related question that is less unclear, viz. the question as to the role of objects called meanings in semantic theory. We especially need to establish whether such an object plays a role in the process of meaningful communication, or whether it is only an artifact of an *account of* meaningful communication. The former possibility is that it is a part of what semantic theory is devised to capture (and hence that in this sense it is "really out there", i.e. really is an object); the latter is that it is an expedient of the theory (and hence it is merely conjured as an object by the theoreticians).

Consider economic theory. The savings possessed by individuals may count as part of its subject matter. Each individual has zero or more savings, and this would most probably influence their economic behavior. Consider, in contrast to this, the purchasing power of individuals. This is not something the individuals have in the sense in which they have the savings; it is rather a characteristic the economists invented to account for ongoing economic situations. So are the meanings like the savings (which their owners have in the literal sense), or rather like the purchasing powers (which they have only within the economists' model)?

The label doctrine opts for the former answer – the link between an expression (sign) and its meaning (signified) is what provides for its meaningfulness, it exists independently of our theories and to elucidate it we need to find out what kinds of entities are *de facto* linked to expressions in this way. The toolbox doctrine, on the other hand, does not recognize any objects that would be part of the *de facto* process of expressions becoming meaningful, they may appear on the scene only when we start to do semantic theory and want to treat the usages of individual expressions as peculiar objects.

Explicit examples are available of semantic theories that treat meanings not as parts of their subject matter but rather as their expedients are those which see meanings as measuring units on a par with meters or kilograms (Davidson, 1977, Churchland, 1979). Dennett (1996, pp. 44-5) writes:

> Propositions, then, are the theoretical entities with which we identify, or measure, beliefs. For two believers to share a belief is, by definition, for them to believe one and the same proposition. What then are propositions? They are, by mutually agreed philosophical convention, the abstract meanings shared by all sentences that ... mean the same thing.

## 7.3 Explication

So what I propose is that the question *Are meanings objects?* (Or: Are they *really* objects?), which is unclear to the point of uselessness, be replaced by the question *Are meanings part of the subject matter of semantic theory, or rather part of its toolbox?* And I want to put forward the answer that the latter is the case, that meanings are *tools* of semantic theory, namely the means of its *explication* of the semantic properties of expressions.

Carnap (1947, p. 8) defines explication as follows:

> The task of making more exact a vague or not quite exact concept used in everyday life or in an earlier stage of scientific or logical development, or rather of replacing it by a newly constructed, more exact concept, belongs among the most important tasks of logical analysis and logical construction. We call this the task of explicating, or of

> giving an explication for, the earlier concept; this earlier concept, or sometimes the term used for it, is called the explicandum; and the new concept, or its term, is called an explicatum of the old one. ... Generally speaking, it is not required that an explicatum have, as nearly as possible, the same meaning as the explicandum; it should, however, correspond to the explicandum in such a way that it can be used instead of the latter.

Quine (1960, p. 238), later, made the idea more precise:

> We fix on the particular functions of the unclear expression that make it worth troubling about, and then devise a substitute, clear and couched in terms to our liking, that fills those functions. Beyond those conditions of partial agreement, dictated by our interests and purposes, any traits of the explicans come under the head of "don't-cares".

Now what I am claiming is that the most adequate way of understanding meanings as set-theoretical objects is understanding them as *explications*, as an encapsulation of the semantic (or pragmatic) properties their expressions have. (Personally, I think that these are mostly use-theoretical, especially inferential, properties,[53] but nothing in this book hangs on such a precise delimitation.)

Note that if we construe meanings thus, we do not have to confront the problem that sets are causally inert. Sets do not enter the process of communication to have causal effects, they just explicate stuff that does. To be meaningful is to have certain semantic properties, such as being employed in a certain way by its users. (And if an expression is standardly used, say, as a warning, then no wonder that its utterance has some direct causal effects.) And what the meaning does is to summarize these, possibly causally effective, properties.

---

[53] See Peregrin (2014).

## 7.4 Meaning and its Formal Explication

How far has this development brought us towards a response to the question *What is meaning?* It is obvious that the interaction of theoretical linguistics and formal logic has been fruitful; but it may also be potentially misleading. Therefore, we must assess its results with due care and keep in mind the kind of achievement they represent.

What I mean by "misleading" is that people working with natural language sometimes take the semantic theories delivered by formal logic simply for granted, as discoveries that are to be disputed no more than, say, the discoveries of physics. And indeed they often *are* discoveries (or incorporate discoveries). However, these discoveries are primarily discoveries of mathematics, and as such are not *directly* relevant for empirical phenomena such as empirical languages. We must not see formal logic as carrying out excavations within a quasi-empirical, metaphysical realm which somehow underlies any human language – what it really does is to study the structures which may be useful for the explication of such languages.

Hence, a mathematician can never tell a linguist anything significant about meaning – he can only supply him with tools. This is not to deny the mathematician's contribution; the tools he delivers are often extremely powerful and the contribution that he, in this indirect way, brings to the study of meaning is therefore formidable. (We have seen that were it not for concepts borrowed from mathematics by Frege, meaning would appear to be utterly unseizable.) Anyway, it is always necessary to distinguish between analyzing the technical tools of doing semantics and the very doing of semantics.

Capturing meanings as set-theoretic objects helps us see the semantic properties of expressions (perhaps – but not necessarily – how they are used in our language games) in a perspicuous manner. Set-theoretical models of language are useful, but they do not vindicate the view that expressions are representations as they are compatible with various use theories of meaning. Indeed, when we look at the inception of formal semantics and examine Frege's maneuver, which is a crucial cornerstone of its edifice, we can see that it is this very idea that fuels it.

Wittgenstein (1953) used to talk about *perspicuous representation* (*übersichtliche Darstellung*).[54] It is an obvious fact that we find some representations (unlike others) of such complex phenomena as language especially perspicuous (Peregrin, 1992). I suggest that explicating meanings as set-theoretic objects by encapsulating the semantic properties of their expressions produces precisely this kind of representation. Hence, what is going on in formal semantics, I think, was nicely summarized by David Lewis (1972, 173): "In order to say what a meaning *is*, we may first ask what a meaning *does* and then find something which does that."

But is it possible to explicate all meanings as set-theoretical objects, so that it would provide for the perspicuous representation? In this book I wanted to reconstruct the movement of formal semantics, initiated at the end of nineteenth century by Gottlob Frege and, precisely in this spirit, brought to full fruition in the last quarter of twentieth century by people like Richard Montague, David Lewis and Max Cresswell.

To grasp the proper point of some of these questions we must see that the entities offered by formal semantics as potential *explicata* for meanings (various kinds of functions, sets, structures) are neither linguistic meanings themselves nor models of linguistic meanings in the sense in which a wooden replica of a ship is a model of the ship. Linguistic meanings are not perceptible things and hence cannot be replicated in this way. We have seen that the reason why Frege proposed identifying the meaning of a predicate with a function from objects to truth values was not that he would have glimpsed such a thing somewhere behind natural language predicates or in the minds of those who use them, but rather because such functions plausibly recapitulated the "semantic behavior" of the predicates. What can be observed and

---

[54] Wittgenstein (1953, §122) states: "A main source of our failure to understand is that we do not command a clear view of the use of our words. – Our grammar is lacking in this sort of perspicuity. A perspicuous presentation produces just that understanding which consists in 'seeing connections'."

what is thus the ultimate data for a theoretician pursuing formal semantics is this "behavior", i.e. the ways in which expressions are employed by their competent users.

# 8 Appendix: the Models

## 8.1 Basic extensional model

1. *Categories of expressions and simple expressions*

1.1 The category of *terms* (**T**) contains an unlimited number of simple terms.

1.2 The category of *predicates* (**P**) contains an unlimited number of simple predicates.

1.3 The category of *propositional operators* (**O**) contains one simple expression, namely ¬.

1.4 The category of *propositional connectives* (**C**) contains three simple expressions, namely ∧, ∨ and →.

1.5 The category of *quantifiers* (**Q**) contains two simple expressions, namely Σ and Π.

1.6 The category of *statements* (**S**) does not contain any simple expressions.

In addition, we assume that we have auxiliary, *syncategorematic* symbols, namely parentheses.

2. *Syntactic rules*

2.1 If $P$ is a predicate and $T$ is a term, then $P(T)$ is a statement.

2.2 If $Q$ is a quantifier and $P$ is a predicate, then $Q(P)$ is a statement.

2.3 If $O$ is a propositional operator and $S$ is a statement, then $OS$ is a statement.

2.4 If $C$ is a propositional connective and $S_1$, $S_2$ are statements, then $(S_1 \, C \, S_2)$ is a statement.

3. *Denotations of simple expressions*

3.1 The denotation of a term is an element of the given universe U.

3.2 The denotation of a predicate is a function from U to B, i.e. the element of [U ⇨ B] (or a subset of U, i.e. a set of individuals).

3.3 The denotation $g$ of a propositional operator is a function from B to B, i.e. the element [B ⇨ B]; the denotation of the operator ¬ is the function given by the table (T¬).

(T¬) $\|\neg\|$ =

    *Tr* ⟶ *Fa*

    *Fa* ⟶ *Tr*

3.4 The denotation of a propositional connective is a function from BxB to B, i.e. the element of [BxB ⇨ B]; the meaning of the operators ∧, ∨ and → are the functions given by the tables (T∧), (T∨) and (T→).

(T∧) $\|\wedge\|$ =

    *Tr, Tr* ⟶ *T*

    *Tr, Fa* ⟶ *Fa*

    *Fa, Tr* ⟶ *Fa*

    *Fa, Fa* ⟶ *Fa*

(T∨) $\|\vee\|$ =

    *Tr, Tr* ⟶ *Tr*

    *Tr, Fa* ⟶ *Tr*

    *Fa, Tr* ⟶ *Tr*

    *Fa, Fa* ⟶ *Fa*

(T→) $\|\rightarrow\|$ =

    *Tr, Tr* ⟶ *Tr*

    *Tr, Fa* ⟶ *Fa*

    *Fa, Tr* ⟶ *Tr*

    *Fa, Fa* ⟶ *Tr*

3.5. The denotation of a quantifier is a function from [U ⇨ B] (sets of individuals) to B, i.e. an element of [[U ⇨ B] ⇨ B] (sets of sets of individuals); the denotation of the quantifier Π is the function that assigns the value *Tr* to a function $f$ just when $f(i) = $ *Tr* for every individual $i \in U$ (or it is such a set of sets of individuals which contains a single set of individuals, namely the one that is identical with U, i.e. contains all the individuals of the universe); the denotation of Σ is the function that assigns the value *Tr* to the function $f$ just when $f(i) = $ *Tr* for at least one individual $i \in U$ (the set of sets of individuals that contains all those sets of individuals that are nonempty, i.e. that contain at least one individual).

3.6. The denotation of a statement is a truth value, i.e. an element of B.

4. *Denotations of compound expressions*

4.1 If $P$ is a predicate and $T$ is a term, then $\|P(T)\| = \|P\|(\|T\|)$.

4.2 If $Q$ is a quantifier and $P$ is a predicate, then $\|Q(P)\| = \|Q\|(\|P\|)$.

4.3 If $O$ is a unary propositional operator and $S$ is a statement, then $\|OS\| = \|O\|(\|S\|)$.

4.4 If $C$ is a binary propositional operator and $S_1$, $S_2$ statements, then $\|S_1 C S_2\| = \|C\|(\|S_1\|, \|S_2\|)$.

## 8.2 Modified extensional model

1. *Categories of expressions and simple expressions*

1.1 The category of *terms* (**T**) contains an unlimited number of simple terms.

1.2 The category of *predicates* (**P**) contains an unlimited number of simple predicates.

1.3 The category of *propositional operators* (**O**) contains one simple expression, namely $\neg$.

1.4 The category of *propositional connectives* (**C**) contains three simple expressions, namely $\wedge$, $\vee$ and $\rightarrow$.

1.5′ The category of quantifiers (**Q**) contains two simple expressions, namely $\exists$ and $\forall$.

1.6 The category of *statements* (**S**) does not contain any simple expressions.

In addition, we assume that we have auxiliary, *syncategorematic* symbols, namely parentheses and an unlimited number of variables $x, y, \ldots$.

2. *Syntactic rules*

2.1 If $P$ is a predicate and $T$ is a term, then $P(T)$ is a statement.

2.2 If $Q$ is a quantifier, $x$ a variable and $T$ a term, and $S$ a statement, then $Qx(S^{T \leftarrow x})$ is a statement. (Recall that $S^{T \leftarrow x}$ denotes the result of replacing $T$ by $x$ in $S$).

2.3 If $O$ is a propositional operator and $S$ is a statement, then $OS$ is a statement.

2.4 If $C$ is a propositional connective and $S_1$, $S_2$ are statements, then $(S_1 \, C \, S_2)$ is a statement.

3. *Denotations of simple expressions*

3.1 The denotation of a term is an element of the given universe U.

3.2 The denotation of a predicate is a function from U to B, i.e. the element of [U ⇨ B] (or a subset of U, i.e. a set of individuals).

3.3 The denotation *g* of a propositional operator is a function from B to B, i.e. the element [B ⇨ B]; the denotation of the operator ¬ is the function given by the above table (T¬).

3.4 The denotation of a propositional connective is a function from BxB to B, i.e. the element of [BxB ⇨ B]; the meaning of the operators ∧, ∨ and → are those functions which are given by the above tables (T∧), (T∨) and (T→).

3.5 The denotation of the quantifier is the element [[U ⇨ B] ⇨ B] (set of sets of individuals); the denotation of the quantifier ∀ is the function that assigns the value ***Tr*** to the function $f$ just when $f(i)$ = ***Tr*** for each individual $i \in U$; the denotation of ∃ is the function that assigns the value ***Tr*** to the function $f$, just when $f(i)$ = ***Tr*** for at least one individual $i \in U$.

3.6. The denotation of the statement is a truth value, i.e. an element of B.

4. *Denotations of compound expressions*

4.1 If *P* is a predicate and *T* is a term, then $\|P(T)\| = \|P\|(\|T\|)$.

4.2 if *Q* is a quantifier, *x* a variable, *T* a term and *S* a statement, then $\|Qx(S^{T \leftarrow x})\|$ is the value of the function $\|Q\|$ applied to the function that assigns the truth value $\|S\|_{\|T\|=i}$ to every element *i* of the universe U (i.e. the value of the function $\|Q\|$ applied to the set $\{i \mid \|S\|_{\|T\|=i} = \textbf{\textit{Tr}}\}$).

4.3 If *O* is a unary propositional operator and *S* is a statement, then $\|OS\| = \|O\|(\|S\|)$.

4.4 If *C* is a binary propositional operator and $S_1$, $S_2$ statements, then $\|S_1 C S_2\| = \|C\|(\|S_1\|, \|S_2\|)$.

## 8.3 Categorial grammar

1. *Categories of expressions and simple expressions*

We have an unlimited number of expressions of each syntactic category, where syntactic categories are given as follows:

1.1. if $C \in CAT$, then C is a syntactic category

1.2. whenever A, B are syntactic categories, B/A is also a syntactic category.

2. *Syntactic rules*

2.1. if $B$ is an expression of the category B/A and if $A$ is an expression of the category A, then $B(A)$ is an expression of the category B.

3. *Denotations of simple expressions*

For each category C we have the set $D_C$ (*domain* of category C) such that if $E$ is an expression of category C, then $\|E\| \in D_C$; where the domains are given as follows:

3.1 If $C \in CAT$, then $D_C$ is some given set

3.2. $D_{B/A} = [D_A \Rightarrow D_B]$.

4. *Denotations of compound expressions*

4.1. $\|B(A)\| = \|B\|(\|A\|)$.

## 8.4 Lambda-categorial grammar

1. *Categories of expressions and simple expressions*

We have an unlimited number of expressions of each syntactic category, where syntactic categories are given as follows:

1.1. if C∈CAT, then C is a syntactic category

1.2. whenever A, B are syntactic categories, B/A is also a syntactic category.

2. *Syntactic rules*

2.1. if $B$ is an expression of a category B/A and if $A$ is an expression of a category A, then $B(A)$ is an expression of a category B.

2.2. if $B$ is an expression of a category B, $A$ is an expression of a category A and $x$ is a variable of a category A, then $\lambda x(B^{A \leftarrow x})$ is an expression of category B/A.

3. *Denotations of simple expressions*

For each category C we have a domain $D_C$ such that if $E$ is an expression of the category C, then $\|E\| \in D_C$; where

3.1 If C∈CAT, then $D_C$ is some given set

3.2. $D_{B/A} = [D_A \Rightarrow D_B]$.

4. *Denotations of compound expressions*

4.1. $\|B(A)\| = \|B\|(\|A\|)$

4.2. $\|\lambda x(B^{A \leftarrow x})\|$ is such function $f$ that $f(d) = \|B\|_{\|A\|=d}$.

## 8.5 Locally intensional logic

1. *Categories of expressions and simple expressions*

We have an unlimited number of simple expressions of each syntactic category, where the syntactic categories are given as follows:

1.1. **T** and **S** are syntactic categories

1.2. whenever A, B are syntactic categories, a syntactic category is also B/A.

1.3. whenever A is a syntactic category, a syntactic category is also A/**I**.

In addition, we assume that we have auxiliary, *syncategorematial symbols*: parentheses, $\lambda$, $^\wedge$, $^\vee$ and an unlimited number of variables for each syntactic category.

2. *Syntactic rules*

2.1. if $B$ is an expression of the category B/A and $A$ an expression of the category A, $B(A)$ is an expression of the category B.

2.2. if $B$ is an expression of the category B, $A$ an expression of the category A and $x$ a variable of category A, $\lambda x(B^{A \leftarrow x})$ is an expression of the category B/A.

2.3. if $A$ is an expression of the category A, then $^\wedge A$ *is* an expression of the category A/**I** .

2.3. if $A$ is an expression of a category A/**I**, then $^\vee A$ is an expression of the category A.

3. *Denotation of simple expressions*

For each category C we have domains $D_C$ and $S_C$ such that $S_C$ is [W $\Rightarrow$ $D_C$] and if $A$ is an expression of the category C, then $\|A\| \in S_C$; where

3.1. $D_T = U$; $D_S = B$ (where U is a given universe)

3.2. $D_{A/B} = [D_B \Rightarrow D_A]$

3.3. $D_{A/I} = [W \Rightarrow D_A]$ (where W is a given set of possible worlds).

## 4. Denotations of compound expressions

If we write $\|A\|^w$ for short instead of $\|A\|(w)$ ($\|A\|^w$ is therefore the value of $\|A\|$ for the possible world $w$), then for every possible world $w \in W$ holds:

4.1. $\|B(A)\|^w = \|B\|^w (\|A\|^w)$

4.2. $\|\lambda x(B^{A \leftarrow x})\|^w$ is the function $f$ that for each $i \in D_A, f(i) = \|B\|\, \|A\|^w = i$

4.3. $\|{^\wedge}A\|^w = \|A\|$

4.4. $\|{^\vee}A\|^w$ is the function $f$ that for each $w' \in W, f(w') = \|A\|^{w'}(w')$.

## 8.6 Globally intensional logic

1. *Categories of expressions and simple expressions*

We have an unlimited number of simple expressions of each syntactic category, where the syntactic categories are given as follows

1.1. **T**, **S** and **I** are syntactic categories

1.2. whenever A, B are syntactic categories, then so is B/A.

In addition, we assume that we have auxiliary, *syncategorematical symbols*: parentheses, $\lambda$, and an unlimited number of variables for each syntactic category.

2. *Syntactic rules*

2.1. if $B$ is an expression of the category B/A and $A$ is an expression of the category A, then $B(A)$ is an expression of the category B.

2.2. if $B$ is an expression of category B, $A$ an expression of category A and $x$ a variable of category A, then $\lambda x(B^{A \leftarrow x})$ is an expression of category B/A.

3. *Denotations of simple expressions*

For each category C we have a domain $D_C$ such that if $A$ is an expression of category C, then $\|A\| \in D_C$; where

3.1. $D_T = U$; $D_V = B$; $D_I = W$ (where U is a given universe and W is a given set of possible worlds)

3.2. $D_{B/A} = [D_A \Rightarrow D_B]$

4. *Denotations of compound expressions*

3.1. $\|B(A)\| = \|B\|(\|A\|)$

3.2. $\|\lambda x(B^{A \leftarrow x})\| = $ is the function $f$ that $f(i) = \|B\|_{\|A\|=i}$

# References

Bar-Hillel, Y. (1953): 'A quasi-arithmetical notation for syntactic description'. *Language* 29(1), 47–58.
Barcan, R. C. (1946): 'A functional calculus of first order based on strict implication'. *The Journal of Symbolic Logic* 11(1), 1–16.
Barwise, J. & Cooper, R. (1981): 'Generalized quantifiers and natural language'. *Linguistics and Philosophy* 4(2), 159–219.
Barwise, J. & Perry, J. (1983): *Situations and attitudes*, Cambridge (Mass.): MIT Press.
Bigelow, J. C. (1978): 'Believing in semantics'. *Linguistics and Philosophy* 2(1), 101–144.
Cantor, G. (1932): *Gesammelte Abhandlungen mathematischen und philosophischen Inhalts* (hg. v. E. Zermelo), Berlin: Springer.
Carnap, R. (1942): *Introduction to semantics*, Cambridge (Mass.): Harvard University Press.
Carnap, R. (1947): *Meaning and necessity*, Chicago: The University of Chicago Press.
Casadio, C. (1988): 'Semantic categories and the development of categorial grammars'. *Categorial grammars and natural language structures* (ed. T. Oehrle et al.), pp. 95–123, Dordrecht: Springer.
Chellas, B. F. (1980): *Modal logic: an introduction*, Cambridge: Cambridge University Press.
Chomsky, N. (1957): *Syntactic structures*, The Hague: Mouton.
Chomsky, N. (1986): *Knowledge of language*, Westport: Praeger.
Chomsky, N. (1993): 'A minimalist program for linguistic theory'. Chomsky: *The view from Building 20*, pp. 1–52, Cambridge (Mass.): MIT Press.
Chomsky, N. (2000): *New horizons in the study of language and mind*, Cambridge: Cambridge University Press.
Church, A. (1940): 'A formulation of the simple theory of types'. *The Journal of Symbolic Logic* 5(2), 56–68.
Church, A. (1956): *Introduction to mathematical logic*, Princeton: Princeton University Press.
Churchland, P. (1979): *Scientific realism and the plasticity of mind*, Cambridge: Cambridge University Press.
Cresswell, M. J. (1973): *Logic and languages*, London: Methuen.

Cresswell, M. J. (1982): 'The autonomy of semantics'. *Processes, beliefs and questions* (ed. S. Peters & E. Saarinen), pp. 69–86, Dordrecht: Reidel.

Cresswell, M. J. (1985): *Structured meanings: The semantics of propositional attitudes*, Cambridge (Mass.): MIT Press.

Davidson, D. (1977): 'The method of truth in metaphysics'. *Midwest studies in philosophy 2: Studies in the philosophy of language* (ed. P. A. French, T. E. Uehling, Jr. & H. Wettstein), Minneapolis: University of Minnesota Press; reprinted in Davidson: *Inquiries into truth and interpretation*, Oxford: Clarendon Press, 1984, pp. 215–226.

Davidson, D. (1990): 'Representation and interpretation'. *Modeling the mind* (ed. K. A. Moyhelin Said et al.), Oxford: Clarendon Press; reprinted in Davidson: *Problems of rationality*, Oxford: Clarendon Press, 2004, pp. 87–100.

Demopoulos, W. (2013): *Logicism and its philosophical legacy*, Cambridge: Cambridge University Press.

Dennett, D. (1996): *Kinds of minds*, New York: Basic Books.

Gabbay D. M. and Woods, J. eds. (2004): *Handbook of the history of logic. Volume 1: Greek, Indian, and Arabic logic*, North Holland: Elsevier.

Dummett, M. (1981a): *Frege: Philosophy of language*, Cambridge (Mass.): Harvard University Press.

Dummett, M. (1981b): *The interpretation of Frege's philosophy*, London: Duckworth.

Dummett, M. (1996): *Origins of analytical philosophy*, Cambridge (Mass.): Harvard University Press.

Egli, U. & von Heusinger, K. (1995): 'The epsilon operator and E-type pronouns'. *Lexical knowledge in the organization of language* (ed. U. Egli et al.), pp. 121–141, Amsterdam: Benjamins.

Frege, G. (1879): *Begriffsschrift*, Halle: Nebert.

Frege, G. (1891): *Funktion und Begriff: Vortrag, gehalten in der Sitzung vom 9. Januar 1891, der Jenaischen Gesellschaft für Medizin und Naturwissenschaft*, Jena: Pohle.

Frege, G. (1892a): 'Über Begriff und Gegenstand'. *Vierteljahrschrift Für Wissenschaftliche Philosophie* 16, 192–205.

Frege, G. (1892b): 'Über Sinn und Bedeutung'. *Zeitschrift Für Philosophie Und Philosophische Kritik* 100(1), 25–50.

Frege, G. (1918): 'Der Gedanke'. *Beiträge Zur Philosophie des Deutschen Idealismus* 2, 58–77.

Frege, G. (1884): *Grundlagen der Arithmetik: eine logisch-mathematische Untersuchung über den Begriff der Zahl*, Breslau: Koebner.

Gallin, D. (1975): *Intensional and higher-order modal logic*, Amsterdam: North-Holland.

Grattan-Guinness, I. (2000): *The search for mathematical roots, 1870-1940: Logics, set theories and the foundations of mathematics from Cantor through Russell to Gödel*, Princeton: Princeton University Press.

Griffiths, O. & Paseau, A. C. (2022): *One true logic: A monist manifesto*, Oxford: Oxford University Press.

Groenendijk, J. & Stokhof, M. (1991): 'Dynamic predicate logic'. *Linguistics and Philosophy* 14(1), 39–100.

Heim, I. (1983): 'File change semantics and the familiarity theory of definitness'. *Meaning, use, and interpretation of language* (ed. R. Bäuerle et al.), pp. 164–189, Berlin: de Gruyter.

Henkin, L. (1949): 'The completeness of the first-order functional calculus'. *The Journal of Symbolic Logic* 14(3), 159–166.

Hilbert, D. (1923): 'Die Logischen Grundlagen der Mathematik'. *Mathematischen Annalen* 88, 151–165.

Hintikka, J. (1975): 'Impossible possible worlds vindicated'. *Journal of Philosophical Logic,* 4(4), 475–84.

Hughes, G. E. & Cresswell, M. J. (1984): *A companion to modal logic*, New York: Methuen.

Janssen, T. M. (1986): *Foundations and applications of Montague grammar*, Amsterdam: Mathematisch Centrum.

Kamp, H. (1981): 'A theory of truth and semantic representation'. *Methods in the Study of Language Representation* (ed. J. Groenendijk, T. M. V. Janssen and M. Stokhof), pp. 277–322, Amsterdam: Mathematisch Centrum.

Kamp, H. & Reyle, U. (1993): *From discourse to logic*, Dordrecht: Kluwer.

Katz, J. J. & Postal, P. (1964): *An integrated theory of linguistic descriptions*, Cambridge (Mass.): MIT Press.

Kemeny, J. G. (1956a): 'A new approach to semantics-part I'. *The Journal of Symbolic Logic* 21(1), 1–27.

Kemeny, J. G. (1956b): 'A new approach to semantics-part II'. *The Journal of Symbolic Logic* 21(2), 149–161.

Kirkham, R. (1995): *Theories of truth: A critical introduction*, Cambridge (Mass.): MIT Press.

Kripke, S. (1963a): 'Semantical analysis of modal logic I (normal modal propositional calculi)'. *Zeitschift für Mathematische Logik und Grundlagen der Mathematik* 9(5-6), 67–96.

Kripke, S. (1963b): 'Semantical considerations on modal logic'. *Acta Philosophica Fennica* 16, 83–94.

Kripke, S. (1965): 'Semantical analysis of intuitionistic logic'. *Formal systems and recursive functions* (ed. J. Crossley and M. Dummett), pp. 92–130, Amsterdam: North-Holland.

Kripke, S. (1972): 'Naming and necessity'. *Semantics of natural language* (ed. D. Davidson & G. Harman), pp. 253–355, Dordrecht: Reidel; later published as a book.

Lakoff, G. (1971): 'On generative semantics'. *Semantics: An interdisciplinary reader in philosophy, linguistics and psychology* (ed. D. D. Steinberg and L. A. Jakobovits), Cambridge: Cambridge University Press.

Lavine, S. (1994): *Understanding the infinite*, Cambridge (Mass.): Harvard University Press.

Lewis, C. I. (1912): 'Implication and the Algebra of Logic'. *Mind* 21(4), 522–531.

Lewis, C. I. & Langford, C. H. (1932): *Symbolic logic*, New York: Century.

Lewis, D. (1972): 'General semantics'. *Semantics of natural language* (ed. D. Davidson & G. Harman), pp. 169–218, Dordrecht: Reidel.

Löbner, S. (1987): 'Natural language and generalized quantifier theory'. *Generalized Quantifiers* (ed. P. Gärdenfors), pp. 181–201, Dordrecht: Reidel.

Mates, B. (1968): 'Leibniz on possible worlds'. *Logic methodology and philosophy of science* (ed. B. van Rootsellar & J. F. Stall), Amsterdam: North Holland.

Meyer-Viol, W. P. M. (1995): *Instantial logic* (dissertation), Amsterdam: ILLC.

Montague, R. (1974): *Formal philosophy: Selected papers of R. Montague*, New Haven: Yale University Press.

Morrill, G. V. (1994): *Type logical grammar: Categorial logic of signs*, Dordrecht: Kluwer.

Oehrle, T. et al., ed. (1988): *Categorial grammars and natural language structures*, Dordrecht: Reidel.
Peregrin, J. (1992): 'Sprache und ihre Formalisierung'. *Deutsche Zeitschrift für Philosophie* 40(3), 237–244.
Peregrin, J. (1994): 'Interpreting formal logic'. *Erkenntnis* 40(1), 5–20.
Peregrin, J., ed. (1999): *Truth and its Nature (if any)*, Dordrecht: Kluwer.
Peregrin, J. (2000a): 'Constructions and concepts'. *Between words and worlds* (ed. T. Childers & J. Palomäki), pp. 34–48, Praha: Filosofia.
Peregrin, J. (2000b): 'Variables in natural language: Where do they come from?'. *Variable-free semantics* (ed. M. Böttner & W. Thümmel), pp. 46–65, Osnabrück: Secol.
Peregrin, J. (2001): *Meaning and structure: Structuralism of (post)analytic philosophers*, Aldershot: Asahgate.
Peregrin, J. (2014): *Inferentialism: Why rules matter*, Basingstoke: Palgrave.
Peregrin, J. (2020): *Philosophy of Logical Systems*, New York: Routledge.
Peters, S. & Westerståhl, D. (2006): *Quantifiers in language and logic*, Oxford: Clarendon Press.
Potter, M. (2004): *Set theory and its philosophy: A critical introduction*, Oxford: Clarendon Press.
Priest, G. (2001): *An introduction to non-classical logics*, Cambridge: Cambridge University Press.
Quine, W. V. O. (1960): *Word and object*, Cambridge (Mass.): MIT Press.
Reichenbach, H. (1947): *Elements of symbolic logic*, New York: Free Press.
Rorty, R. (1979): *Philosophy and the mirror of nature*, Princeton: Princeton University Press.
Russell, B. (1905): 'On denoting'. *Mind* 14(56), 479–493.
Russell, B. (1908): 'Mathematical logic as based on the theory of types'. *American Journal of Mathematics* 30(3), 222–262.
Sellars, W. (1974): 'Meaning as functional classification'. *Synthese* 27(3-4), 417–437.
Sgall, P., Hajičová, E. & Panevová, J. (1986): *The meaning of the sentence in its semantic and pragmatic aspects*, Prague: Academia.

Slater, B. (1991): 'The epsilon calculus and its applications'. *Grazer Philosophische Studien* 41, 175–205.

Stalnaker, R. (1986): 'Possible worlds and situations'. *Journal of Philosophical Logic* 15(1), 109–123.

Tarski, A. (1935): 'Der Wahrheitsbegriff in den formalisierten Sprachen'. *Studia Philosophica* 1, 261–405.

Tarski, A. (1936): 'Grundlegung der wissenschaftlichen Semantik'. *Actes du Congrès international de philosophie scientifique* 3, pp. 1–8.

Tarski, A. (1944): 'The semantic conception of truth'. *Philosophy and Phenomenological Research* 4(3), 341–375.

Thomason, R. (1980): 'A model theory for propositional attitudes'. *Linguistics and Philosophy* 4(1), 47–70.

Tichý, P. (1971): 'An approach to intensional analysis'. *Noûs* 5(3), 273–297.

Tichý, P. (1975): 'What do we talk about?' *Philosophy of Science* 42(1), 80–93.

Tichý, P. (1978): 'De dicto and de re'. *Philosophia* 8(1), 1–16.

Tichý, P. (1986): 'Constructions'. *Philosophy of Science* 53(4), 514–534.

Tichý, P. (1988): *The foundations of Frege's logic*, Berlin: de Gruyter.

Van Benthem, J. (1984): 'Questions about quantifiers'. *Journal of Symbolic Logic* 49(2), 443–466.

Van Benthem, J. (1988): 'The Lambek calculus'. *Categorial grammars and natural language structures* (ed. R. T. Oehrle, E. Bach & D. Wheeler), Dordrecht: Springer.

Van Benthem, J. (1997): *Exploring logical dynamics*, Stanford: CSLI.

Van Benthem, J. & ter Meulen, A., eds. (1996): *Handbook of logic and language*, Amsterdam: North Holland.

Werning, M., Machery, E., and Schurz, G., eds. (2005): *The compositionality of concepts and meanings*, Frankfurt: Ontos.

Wittgenstein, L. (1922): *Tractatus logico-philosophicus*, London: Routledge

Wittgenstein, L. (1953): *Philosophische Untersuchungen*, Oxford: Blackwell.

Zimmermann, T. E. (1989): 'Intensional logic and two-sorted type theory'. *The Journal of Symbolic Logic* 54(1), 65–77.

www.ingramcontent.com/pod-product-compliance
Lightning Source LLC
Chambersburg PA
CBHW050758160426
43192CB00010B/1561